# DEMYSTIFYING PROJECT CONTROL

DIMITRIS ANTONIADIS

ISBN: 9781718063099

This book is dedicated to my wife for her immense support and devotion throughout my entire career

# ACKNOWLEDGEMENTS

This book is the culmination of 30 years' experience in the field of project management. Throughout this period I was influenced by and hopefully have influenced a few individuals. I will be forever thankful for Ken Thompson who back in 1988 brought me into the profession and employed me as Project Control Engineer at Balfour Beatty Projects and Engineering. From that moment on a number of individuals shaped my knowledge and professional capabilities. Some of those influential individuals are Mike Brown (retired now from Rolls Royce), Andrew Rogerson (at Southern Water), Rod Capper (Kellogs Brown & Root), Jerry England and David Glendinning (Thames Water Engineering) and Russell Batchelor (BAA – HET). There were also all those that worked with me in my departments and the projects and programmes of works I managed. They made me think, develop, challenge and change and I owe them an incredibly massive thank you. When they read this book they will recognise the points I make.

Two individuals I would like to thank for their encouragement during the writing of this book are Dr Edum-Fotwe of Loughborough University and Jonathan Norman Knowledge Hub Manager at the Major Projects Association. I would also like to thank the proof-readers of this book, for all the patience, literary points of view and long hours of work and effort, my daughters, Cathryn and Joanna.

My deepest gratitude goes to my wife and daughters for their support.

Finally I would like to address all those who have been questioning the reason for carrying out such a task at this late stage in my life. The simple truth is that, in addition to my PhD on Complexity in Projects, my main driver has been leaving behind a legacy and also because I have always lived my life by this maxim:
"*See what no one else sees*"; "*See what everyone else chooses not to see out of fear, conformity and laziness*" "*See the whole world anew each day*".
From the 1998 film and true story of Patch Adams.

# CONTENTS

## LIST OF TABLES



## LIST OF FIGURES

(*) Author's note

The author wishes to apologise to the readers as some figures, due to resizing, are not as clear as the original. If you would like to discuss any of the figures the author can be contacted on LinkedIn or via his website.

# LIST OF ABBREVIATIONS

| Abbreviations | Description |
|---|---|
| AFP | Application For Payment |
| APM | Association for Project Management |
| BIM | Building Information Modelling |
| BoK | Book or Body of Knowledge |
| CAD | Computer Aided Design |
| CBS | Cost Breakdown structure |
| DBS | Document Breakdown structure |
| ERP | Enterprise Resource Planning |
| EVA | Earned Value Analysis |
| IP | Intellectual Property |
| KM | Knowledge Management |
| LLs | Lessons Learnt |
| LPMO | Labour, Plant, Material, Other |
| MS | Microsoft |
| OBS | Organisational Breakdown structure |
| PBS | Project / Product Breakdown structure |
| PC | Project Control |
| PCP | Project Control Plan or Project Control Handbook |
| PEP | Project or Programme Execution Plan |
| PM | Project Manager |
| PMI | Project Management Institute |
| PMO | Programme Management Office |
| PMP | Project Management Plan |
| PQP | Project Quality Plan |
| PgBS | Programme Breakdown structure |
| QS | Quantity Surveyor |
| RACI | Responsible, Accountable, Consulted, Informed |
| RAM | Responsibility Assignment Matrix |
| RBS | Risk Breakdown structure |
| WBS | Work Breakdown structure |
| WPEP | Work Package Execution Plan |

# ABOUT THE AUTHOR

Dimitris Antoniadis is a consultant providing support to organisations on Programme Management, Programme Management Office (PMO), Transformation and other relevant operations. He has more than 30 years' experience in the field of Programme and Project Management and has covered all project phases from concept to handover and operation / maintenance.

He has held Senior Management posts in major utilities, infrastructure and construction organisations delivering programmes of works such as, Egnatia Odos (£3.2Bn), Southern Water Capital programme (c£1.5Bn), UK Power Networks (c£1.1Bn), Thames Water Engineering Capital programme (c£1Bn), BAA-Heathrow East Terminal (£1.4Bn), Nottingham Building Schools for the Future (£250M).

He has set up and managed a number of Programme Management Offices, delivering a wide range of responsibilities, including responding to Parliamentary questions on a multi-billion budget programme of works.

As Head of Programme Management Office (PMO) he has led / co-led major transformation programmes and has integrated client and contractor PMO software systems for Alliance working environments. The PMOs played central role in covering interfaces, upstream to Executive Management and other Directorates, as well as downstream to delivery by Alliance organisations.

He is a visiting lecturer at a number of UK Universities.

He has presented, as a guest speaker, at international conferences and published a number of papers on project management, PMO and complexity topics in major industry journals. He has also contributed two chapters to the book '*Chaos and Complexity Theory in Management*' (ISBN 978-1-4666-2509-9) and a chapter to the International Centre for Complex Project Management '*Global Perspectives and the Strategic Agenda to 2025*' (ABN 77 12 367 373).

He was awarded the PhD, from Loughborough University, for research on the subject of 'Managing Complexity in Project Teams'.

# 1. PREFACE

*It is an early Saturday morning in April 2010 and I am standing outside the University of West London classroom where I will deliver the first (of 13) lectures on the part time course of Project Management for the BSc students of Business Studies.*

*The students that are also waiting outside the room don't know who I am and pass their time discussing the other subjects and coursework. They have all just finished their working week and Saturday is the only time they come together to discuss coursework. The subject changes and they start talking about the textbooks proposed on Project Management and how, since they are practitioners, they are frustrated that these do not include actual real life examples. Hearing that I cannot help but smile because of what I have prepared for them. Each one of my lectures, and the accompanying material I have prepared, is based on real life examples that I had gathered from all my years in the industry.*

*My aim was to take them through what is really happening outside the classroom and in the real world.*

*That is probably when the seed of an idea to write a book on Project Control based on real life examples was planted in my mind. I imagine my brain has been working in the background to build the content and structure of what you see now in front of you.*

The following pages are a culmination of my 30 years' experience in the industry. I hope this will help you in understanding and demystifying things about Project Control.

'Demystifying project control'

Where to start on this topic?

Project control is mentioned in all project management literature and in literature from the two main project management professional bodies, the Association for Project Management (APM) and Project Management Institute (PMI). Yet it still has not been properly defined.

After 30 years in the profession, I decided to take up the challenge of demystifying project control, with the aim of helping to contribute towards this knowledge area within the project management field.

I am tempted to start with a definition to help you frame the subject. The trouble is, there is no straightforward definition.

The APM Knowledge guide: 'Introduction to Project Control' (2006) gives two definitions:
'*The application of processes to measure project performance against the project plan, to enable variances to be identified and corrected, so that project objectives are achieved.*'

The second and broader one is:
'*The application of project, programme and portfolio management processes within a framework of project management governance to enable an organisation to do the right projects and to do them right.*'

The guide proceeds to list the processes that are included within project control and give a brief description of what is involved:

- *Controlling the scope of the project – controlling change;*
- *Ensuring that the project's products/deliverables fulfil their requirements – controlling quality;*
- *Ensuring that activities happen on time – scheduling;*
- *Ensuring that work is performed within budget – cost control;*
- *Managing risks;*
- *Managing problems and identifying issues (and obtaining external help to resolve them);*
- *Making sure that the project leads to benefits for the organisation.*

'*The control processes involve the collection and analysis of data, the identification of trends and variances, forecasting and the reporting of progress. It is also essential that the information gathered is acted on – without effective responses to actual and potential problems, the project is merely monitored, not controlled.*' (APM, 2006)

Some points that are missing or are not clear from the above are:

- Estimating project costs;
- Document control;
- Integration of the project control processes.

So, despite its importance in the world of managing projects Project Control partly remains a mystery. There are limited publications with practical examples that describe what it is, how it should be structured, performed, the software systems used, what are the issues and generally the wider requirements. Some individual parts of project control have been included in books about Project Management but to my experience there are no specific publications on project control as a whole.

The importance of project control can be seen below in Figure 1.1. This is the result from a survey that I conducted with 100 professionals from six organisations – Client and Contractor (Antoniadis, 2009) during my research on complexity in project management. The histogram indicates that after the Project Manager's Management / Leadership style the second highest factor that is thought to affect the project management success is Monitoring and Control.

Figure 1.1. *The histogram shows the level of importance practitioners place on monitoring and control for the success of the quality of project management.*

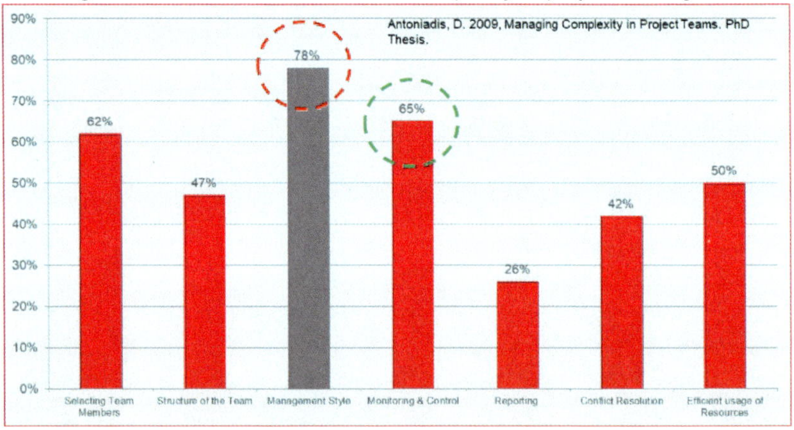


Contribution of project management sub-processes to the success of the quality of the project management for levels – Substantial to Excellent

Perhaps there are reasons that project control has not attracted the attention of authors and books on the subject have not been published. I will touch on some of the issues that make it difficult to write about project control in the following chapters.

I must say that, from my experience, it is not easy to find examples from practice or practical implementation, apart from those that have been given at conferences and the various PMO or Project Control events. However, these examples / outputs are not readily available.

In this book I set out to explore these issues and will propose, based on my experience in the subject, answers and the best approach to take. I will not duplicate some of the excellent work done by the Association for Project

Management (APM), the relevant APM SIGs or the Project Management Institute (PMI). Instead, in certain chapters I will direct the reader to the relevant publication by APM or PMI for more detailed definitions.

This book is aimed for both practitioners and project management academics. However, as it is not aimed to be a review of theory, it will not concentrate on extensive theoretical background but will capture briefly some of the latest thinking from academic / professional journals. This will be covered in the next chapter. In all other chapters I will be covering project control from the practical side.

Also, as the majority of my 30+ years of experience has been in UK I will cover the practices established in UK. Some, however, are replicated in other countries as well. Some of the examples I will be presenting I have implemented whilst working abroad with an international contractor.

Finally, I have included a chapter at the end of the book, which I called '*Exercises*'. I have set up seven areas / exercises that will enable you to embed what you have read and perhaps to prompt you to research and read further on the subject, as part of your continuous learning process. Please do not underestimate their importance, if nothing else they will help you apply what you have read as well as consider the challenges you face in your working environment.

# 2. THE BASICS FROM THEORY

Project control, at least in the UK, is seen as a support function to the Project Manager.
Ning, Y (2017) states that according to Leifer and Mills (1996:117) control is defined as *'a regulatory process by which the elements of a system are made more predictable through the establishment of standards in the pursuit of some desired objective or state'*.

Theoretically, organisations choose their control strategy based on two theories, the agency or the transaction cost economics theories (Ning, 2017). The former depends on the project characteristics, how easily tasks can be programmed and what outputs are measured by the Controller's knowledge. The latter also looks at the transaction cost of project characteristics but in particular asset specificity, frequency of transaction and uncertainty.

According to theory there are two main control mechanisms, the formal and the social.

Formal mechanisms follow a mechanistic approach dependant on contractual completeness. It has rules for each party, based on the contract, and works through surveillance and comparison between agreement and outcomes. However, it ignores interpersonal behaviours. There are two main modes, outcome and behaviour control (Das & Tent, 2001). The former is based on specifying outcomes to be monitored and relevant rewards. The latter specifies how controlees should act through procedures, various routines and rules. All, or at least all organisations, client and contractor (controller and controlee) have a clear view of their formal and behaviour control requirements.

Social control, takes into consideration joint decision making and problem solving which is built from relationships and informal control tools. It promotes trust and mutual gain (Rowley et al, 2000; Larson, 1992). The social control category requirements, which apply more to a partnering or collaborative environment, are also identified and are seen to be implemented to various degrees.

However, let's discuss what the industry sees as the objective of project control.

1. The Objective and Defining Project Control.

The objective of project control is to enable informed decision making by providing integrated and accurately analysed information on time, cost, performance and risk to all project stakeholders as required.

Figure 2.1 below describes the practical application of project control.

Figure 2.1. *The project control loop.*

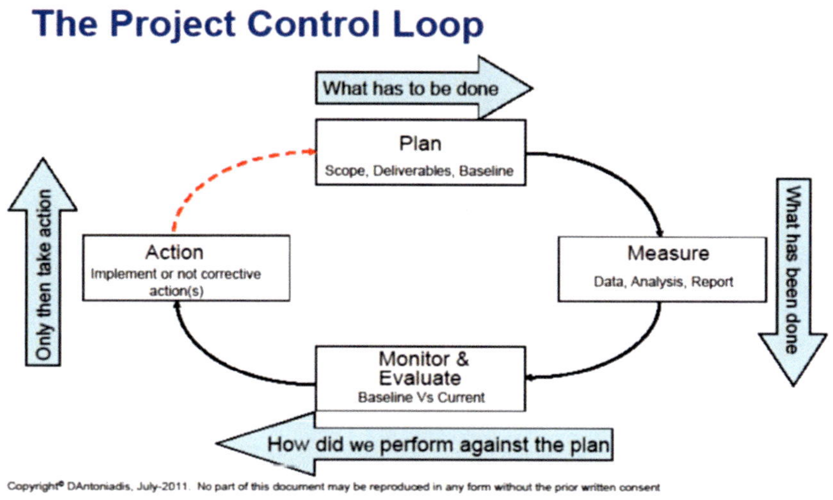


Taking Figure 2.1 and the questions asked into account, a single minded and simplistic view point is that, project control provides an oversight of the project's status makes recommendations and provides forecast information to get the project back on track. It allows for projects and organisations to review project performance at specific intervals and then according to contractual terms and conditions agree payments, etc.

In Figure 1.1, I gave an indication of the practitioners' response to the importance of 'monitoring and control'. However, we must consider how project control adds value to the delivery of the project.

Figure 2.2 below captures the majority of points about the value project control adds to the overall process of project management. It was used for a presentation I gave to practitioners of a major infrastructure organisation. The significant point made at the time was that project control acts also as an integrator, enabling interfaces and communication. The bottom line is that project control is not all about '*Control*'. The points shown are not restrictive and I am sure further benefits could be included. It is important that these are highlighted to practitioners so they are informed and understand the advantages.

What one needs to have in mind is that the majority of the tasks that are to be performed by project control (department or individuals) are a result of / an output from the use of software tools that support the profession. Therefore, and depending on company decisions on software tools, project control objectives will either have limitations or there will be more flexibility.

Figure 2.2: *How project control adds value.*

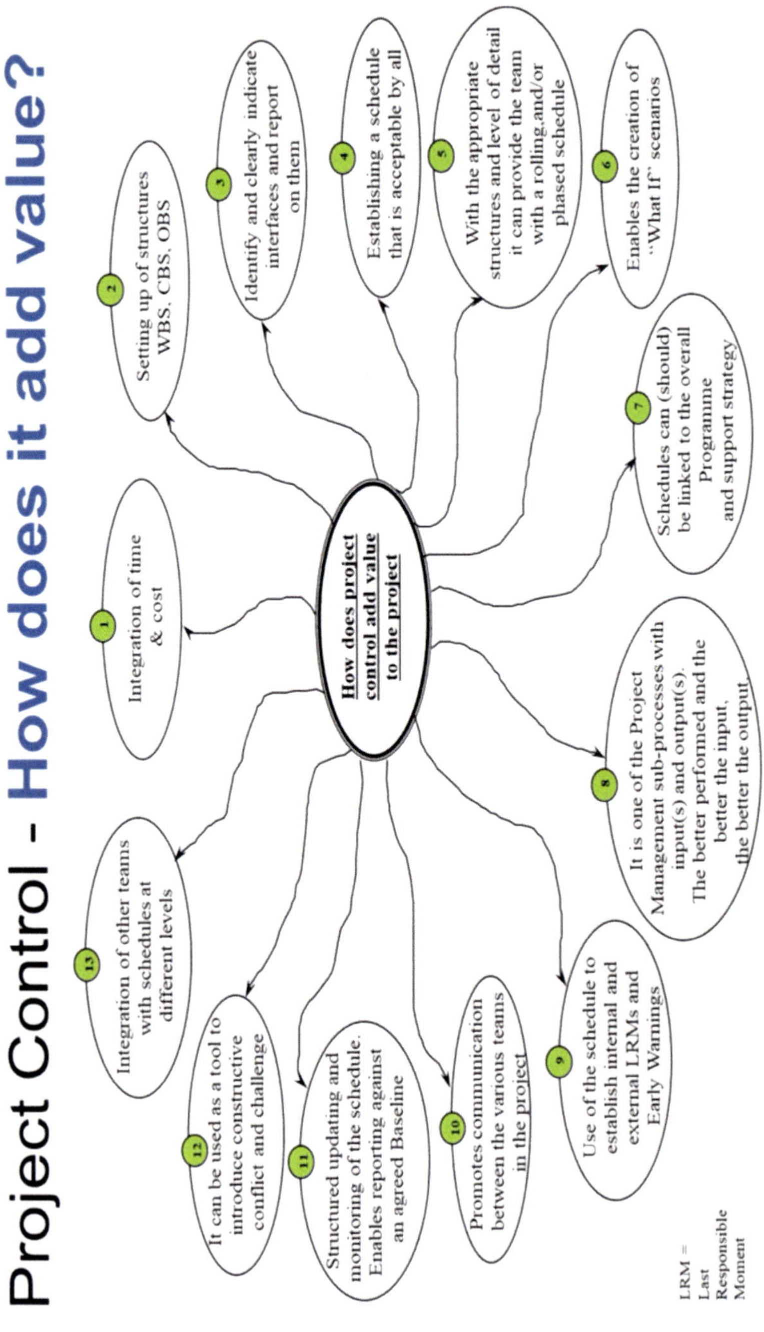

2. The reality and some basics

In reality, despite the fact that various agreements may be in place, including on the use of a common software tool, information asymmetry can be caused by different perceptions of project control between individuals or the controller's and controlee's corporate requirements.

A simple example of this is where in a partnering environment, with common tools and systems, the controller follows a monthly progress update process whereas the controlee's corporate requirements are for weekly progress updates and only focusing on the time element. Of course, I can hear you thinking "they could do the updates offline". Well, in a partnering environment where the organisations are using a common planning tool to do their cost loaded programmes this means taking the programme (the whole or part of it) offline, with possibly other organisations' information, updating the relevant part, extracting the data and then feeding these back into their corporate tools.

*Is this realistically possible?*
*Doesn't this mean duplication of effort and therefore going against the benefits of partnering?*
We will examine some answers to these questions in Chapters 7, 8 and 9 and the various sections that will follow.

So, even with a predefined (contractual) control strategy, project control causes control issues by this information asymmetry, and this asymmetry does not take into account the complexity caused by the interconnections. We will see more on the complexity issues raised in the following sections.

Here are some questions raised by the information asymmetry:

- *Can it be minimised or at least managed?*
- *Can individuals at various hierarchical levels understand the information asymmetry and adhere to guidelines?*
- *Is there ever an end to project control issues?*
- *Do the formal and social control categories clash in the case of a collaborative environment?*
- *Despite all the good will can the unsurmountable obstacles be understood by the higher echelons of the organisations?*

Responses to the above questions will be given as we go through the book. However, some of the answers are dependent on the country / continent where project control is practiced since the weight and responsibility given to project control varies. As this book focuses on Project Control (PC) practices, levels of organisational responsibilities and corporate decisions, regarding hierarchical structures will not be addressed. I will only describe the structure and responsibilities within the department.

Another factor that affects the PC tasks, from the side of the organisation delivering the project – controller or controlee – is the number and size (budget) of projects delivered. On large projects the majority of PC tasks, which will be described later on, are normally performed by dedicated PC personnel. On small projects several of these tasks are performed by the Project Manager (PM).

When considering project control there is also the point about the type of organisation – Client, Contractor, Consultancy. The approach and scale of implementation of project control varies between the former two types. The latter, consultancy, essentially provides the services required, which is services that have been won as a commission from the relevant organisation.

Despite the variances, which are and should be minor in terms of implementing generic project control, I will not be distinguishing these further. Principles are principles and need to be followed irrespective of whom by and in what capacity they are applied.
Having briefly covered some theoretical and peripheral points, let's see what project control is, as in what should be covered by project control.

The project control department, if it exists, has the responsibility to gather and report on cost, time, risks and other project information generated from the implementation of the various control processes (planning, cost management, change management, progress monitoring, resource management, risk & opportunities management, document control, etc.). Where it doesn't exist as a department then these tasks are carried out by the PM or others as delegated.

Activities within a project that are covered by project control vary. However, some of the basic ones are:

3. The Project Control functions

1. Planning / Scheduling:

Although these are considered as two separate activities and processes, in reality in most organisations they are performed by one person in collaboration with the other project team members. Having said that, there is an impression (from individuals in the industry) that the planner/project controller does not really need to speak to the team and that they should be drawing programmes / schedules without contribution from the team. This is not something I agree with, and as it has been witnessed on projects, this causes unnecessary tension. (*For those of you practitioners that are reading this book now you know exactly what I mean*).

In this book planning / scheduling will be considered as the process of linking logically resourced activities (as in cost or person hours), identifying the approach and the critical path to deliver the project scope.

The APM Planning, Scheduling, Monitoring and Control document (APM, 2015) gives the definition of the two processes as follows:

Scheduling: '*usually involves the input of the plan into a scheduling software. It involves the calculation of activity durations required; it defines the logic and sequence and the calculation of critical path, float and start and finish dates of individual activities, and thus determines the feasibility of delivering the project within the desired completion dates and budget*'.

Planning: '*is about identifying at an early stage the means, resources, and actions necessary to accomplice an objective*' (APM A guide to conducting Integrated Baseline Reviews, 2016).

2. Resource Management:

Resource management is concerned with the efficient and effective utilisation of an organisation's resources by attempting to balance three variables – 'Capacity', 'Demand' and 'Supply'. As far as the project is concerned, it is about the uninterrupted and effective utilisation and management of resources. (APM, 2015)

3. Cost Control (including change management):

The APM guidance (2015) gives the definition of the process as follows: '*It is the process of collecting actual costs and collating them in a format to allow comparison with project budgets, identifying variances to inform decision making and allow action to be taken. It includes cost forecasting*'. The formal process, based on the type of the contract that allows the raising of, assessment and approval of change requests is defined as change management / control.

4. Estimation:

The APM guidance (2015) gives the definition of the process as follows: '*Estimating is the process used to quantify the cost of services, materials and resources required to deliver the* project'.

There are two main methods, the approximate and the definitive.

Within the 'approximate' method we have three types of estimating: 'specific analogy estimating', 'parametric' and 'Delfi technique'. Within the 'definitive' method we have, four types of estimating: 'detailed', 'activity based costing' (ABC), 'task analysis' and 'Level of Effort' (LoE). (Further details on the relevant methods can be found in APM guidance documents or relevant books).

5. Performance Reporting:

The APM guidance (2015) gives the definition of the process as follows: '*It is the measurement of progress and spending made to date compared to that of the planned'*. Reporting, and that is not just of performance, enables management to make informed decisions. This is done based on the information provided at that particular point in time and also by looking ahead, for example six to 12 weeks forward. The APM document also provides some advice on the Project Performance Indicators and Key Performance Indicators that can be used.

6. Document Control / Management:

The APM guidance (2015) gives the definition of the process as follows: '*It is the collection, storage, dissemination and archiving of documentation in a structured manner*'. The process of maintaining the document records and the tracing of documentation flow with the appropriate version and revision control is called document control.

7. Risks, Opportunities and Issues:

The APM guidance (2015) gives the definition of the process as follows: '*It is the management of threats and opportunities*' that enable the project team to ensure the best outcome. The various tasks that are required to be undertaken comprise the risk management process. There are basically two types of risks analysis that can be undertaken, quantitative (based on statistical – Monte Carlo – analysis) and qualitative.

8. Project Assurance:

The wider concept of project assurance is described, by APM, as the process that enables the delivery of an objective and independent oversight of the project future performance. Performance is measured in terms of achieving the time, cost, quality and scope objectives as well as delivering the benefits.

The integrated output of the above activities is considered as project control.

The reason for considering these as activities, rather than focusing on roles, is that sometimes the same activities are carried out either by one or two people. For example, a project planner may carry out the cost loading of a logically linked programme and perform site progress review and cost evaluations. Therefore one person conducts three activities. On other occasions, these activities are performed by two or more individuals, e.g. planner, quantity surveyor, site engineer.

All the above terms that describe processes and activities within project control give the impression that everything is about control. However, the successful implementation of project control is highly dependent on how a number of soft issues are addressed. These will be described in detail later on in this book.

I have presented brief definitions for the functions considered as PC, however, there is one activity that could vary significantly depending on the context and it is usually the one that is not managed properly. This activity is Planning / Scheduling (from now on referred to as 'planning').

Theoretically, planning is about the logically linked programme that describes, at a particular point in time, what needs to be done to deliver the project / specific works / work package. Implementation varies from the simple logically linked programmes providing critical path, etc. to fully cost loaded and logically linked programmes that combine time and cost.

In Chapters 4 and 5 we will identify the potential challenges faced as a result of the choices made in approaching planning.

It is important for planning to be given its due time and attention. Planning is not effective if it only involves colouring MS Excel cells or drawing text boxes in MS Power Point. They generate fictitious Critical paths (since no forward / backward pass has been calculated) and don't present a programme or project delivery plan that can be relied upon. In Chapters 8 and 9 we'll see the 'motives' behind these pseudo-programmes and why well-seasoned professionals do this.

In this book the notion of drafting pseudo-programmes will not be considered. However, this had to be mentioned as it is something that is practiced, even by senior management staff, with no comprehension of the problems caused up as well down the line.

In this book, I will use the fully logically linked and cost loaded programme as the basis from which to explore the aspects of project control.

On many occasions project control is confused with that of reporting and is considered only as a function that extracts and manipulates information from relevant tools. I can hear some of you readers thinking "yes, if they have to report by exception on variances or Earned Value indices". But, here come the interesting questions:

- Do you call this project control or reporting by exception which can be done by anyone that knows about the project?
- Who knows more about the project, the PM or a project controller? Who is accountable for the project?
- Why would someone, who is not part of the project, be reporting on variances without the PM's knowledge?
- Is this project control?

The project controller, using the appropriate methods, interrogates the data received, enters information in the software tools, examines the output and presents this for discussion, review and approval to the PM, who is accountable for the project. The PM owns what is reported upstream and can describe why it has happened and the rationale. Whereas supporting team members own the accuracy and are responsible for providing advice on 'the cause(s)', 'the concerns' and from the look-aheads 'the mitigation actions that can be taken'.

Project control as a group of processes, nowadays, is heavily supported by software tools. Those of you that were part of the industry in the mid '80s / early 90s will remember all the manual means of keeping track of project information. Those days even MS Excel was not well developed and various means were used as tools to support the PMs, not to mention the 'on the arrow' planning method.

Today, we have software that integrates information from various tools using the Enterprise Resource Planning (ERP) approach, and rolls-up data for global reporting. We are now looking at the function of project control as one that has to interact with other parties, external to the project team, through the information received and produced. For example, a few years ago one could only theoretically speak of the programme of works informing an asset registration system. Now, it is expected that as soon as a milestone or a resourced activity is complete, the planning software automatically informs the asset registration software and this kicks off the other relevant processes. In all cases, as the basic theory of complexity sets out, interfaces cause complexity.

Later on, in Chapter 7, we will discuss in more detail how information needs to be integrated and what is required. Then, in Chapter 9, we will look at complexity and its effects on the reporting process.

The important point to stress from the outset is the required integration of information. If this is considered by the teams and organisations as an expanded part of project control then it can deliver major benefits down the road not only to the project but to the organisation, as well as those teams and organisations that support the project.

In order to achieve this integration, a number of structures have to be set up. This starts with the standard project management ones. The one structure that has to be used as the core is the project Work Breakdown Structure (WBS). We will discuss in more detail in Chapter 3 what structures have to be used as a minimum, what information these have to carry and how these can be organised to integrate project control as well as other information for departments such as Finance, Asset Management and others.

By setting up a central software tool and standardising project structures we enable the rolling up of information from Projects to Programmes, make reporting to the appropriate levels easier and help to retain the PM's accountability. However, the minute information leaves the system and is reported through other means we cannot hold people accountable.

It is only with the alignment of all the above and standardisation that we are able to integrate information and generate reports. Therefore, reporting is an output of a good project control and NOT project control itself.

Systems and structures allow for the flexibility in reporting, as long as this stems from the software tools that hold the information. However, this flexibility has been misunderstood and reporting requirements change freely and frequently. This is used as an excuse to transfer information off the tools that generate / hold that information; to alter the information exported; to add other / '*explanatory*' information and then present it. This then affects the output and creates the usual 'data integrity' question.

This chapter aimed to cover a lot of ground and some of the basics of project control, though not in extensive detail. For further reading refer to the references made and the bibliography.

In the following chapter I will look into and describe these standard structures. Together with the definitions, given by APM, I will present examples from practice.

# 3. THE PROJECT STRUCTURES

The project structures used by project control teams will be described in this chapter. However, apart from giving a very brief description of the standard definitions and examples, we will look at other important factors and points that you need to know about the structures. Of course these will highlight the importance of and the reasons for the structures, as well as allow you to look at them from a different perspective.
We will look at the primary ones and the supplementary ones that are rarely mentioned in the standard text books. We will also consider the outputs from combining structures, some of those are well-known and others not so well-known.

So, amongst other points, in this chapter I will address the questions, why we need to use structures, if the structures can be flexible and how the structures enable project control to deal with the effects of complexity.

The structures we will talk about are:

- The Work Breakdown Structure (WBS);
- The Organisational Breakdown Structure (OBS);
- The Cost Breakdown Structure (CBS);
  This will include the Resource Breakdown for both, the people hours and cost.
- The Risk Breakdown Structure (RBS);
  The RBS is used mainly, but not exclusively, in Client organisations where various risks are rolling up to asset levels and in many cases are combined with other Company risks;
- The Document Breakdown Structure (DBS);
  Although this is not a well-known structure, it is one that runs in the background and supports Document Control, Information Management, Building Information Modelling (BIM) and other processes indirectly involved with project control.
- The Project / Product Breakdown Structure (PBS).
  Describes what products the project should produce. Some organisations use this to assemble packages of components from different suppliers/manufacturers.

In addition to the above, we will cover another structure that is used to support project control and the product of combining two primary structures to produce a very helpful matrix. These are:

- The Programme Breakdown Structure (PgBS);
- The Responsibility Assignment Matrix (RAM), which plays a very important role.

1. 'WHY' we need to use structures

Once the scope and the deliverables of a new project are set out, the team and its leader are tasked to deliver the outputs and measure these within certain parameters. From the early days and until now we have come up with various parameters to measure the successful delivery of a project. Two of these groups of parameters are:

- The Golden triangle – Cost, Time and Quality
  and a very interesting article on this has been written by Atkinson (1999) with the title 'Cost, Time and Quality; or two best guesses and a phenomenon.
- The Time, Budget, Function, Quality, Conflict (Hughes, 1998)
  - Time, representing and monitoring the time element of the project.
  - Budget, parameter that monitors the Cost & Finance.
  - Function, parameter for monitoring what is to be built.
  - Quality, parameter for monitoring the degree of excellence.
  - Conflict, parameter for monitoring the control of conflict.

In order to decide how the deliverables are best measured, we first need a way to structure things and understand how to perform the relevant actions. If this is ignored then the risk is not only that we do not deliver the required output but that the task is not delivered efficiently and effectively.

When the theory of project management was developed it was considered appropriate to structure the deliverable(s) into packages of work as well as to understand who would deliver them. So, this led to the creation of the Work (What) and the Organisation (Who) Breakdown Structures.

So in considering the above we might first ask, 'What am I to deliver' and 'What are the work packages that will deliver the project'. This way we structure the WBS. We form a view and to help us do this we have a number of means. What counts is the fact that our brains think in structures. After the first step is taken – and remember 'a journey of a 1000 steps begins with the first step' (Lao Tzu) – and in order to complete the WBS, you can always get the project team involved either through a brain storming session or by communicating to them what has been set up and requesting input. Whatever the means, and as this is one of the primary structures which should be led by project control, the WBS has to be owned by the whole team. The team has to understand what needs to be delivered.

By setting up a proper WBS, as part of the Project Management Plan (PMP), we are able to minimise future problems. Busby and Hughes (2004) established that errors at the early stages of projects instil '*pathogens*' which then at a future point become more profound and cause major disruptions. They called this '*pathogens and incubation period*'. By setting up the WBS properly from the early stages of the project you have partially mitigated the effects of the relevant pathogens that could affect the project at a later stage. This is done by giving clarity to the scope, the deliverables and the content of work; establishing assumptions made; identifying the interfaces and the project phases to mention but a few. We will explore further down how the structures allow this.

Having set up the WBS and depending on experience, we either think about 'Who' will deliver the packages – OBS – or 'How much' each package will cost – CBS – so that we can establish the estimated project cost.

Looking at the OBS first, and 'Who' will deliver the work packages, is natural having established the 'what'. So, again we enter names, departments, or companies and either enter these next to the WBS elements or in a separate structure. How we do this matters less than the actual setting up of these structures.

Although the OBS, and other structures, has been around for a few decades there is a new proposal which considers pulling together the structures in a 3D view (Antoniadis, 2013) but it is still in early development stages. As is the case with the OBS, which involves people, often different levels of hierarchy are required to engage in projects and this is not supported well by the current approach. Whereas, when we start looking at things in 3D we can more easily bring together different levels of organisational structure and present how these are linked.

Following on from setting the WBS ('What') and the OBS ('Who') we need to start considering the project costs. Let's not forget at this point that we will also need to be working on the logically linked project programme. This is where the CBS ('How much') comes in. The interesting point about it is that the structure has to be set up centrally, at company level, as a number of other processes, e.g. invoicing, month-end financial reports etc., are also dependant on the way project costs are captured and rolled up. The main reason supporting the use of the CBS is that, not only does it serve the grouping of project costs and rolls this up according to the type, but it is also used centrally in order to pay invoices, salaries, via the use of timesheets, and in combination with other structures enables the set-up of shareholders' reports.

Having considered the above, we have established why setting up these structures at the initial stages of the project is paramount. It does not take long to spend some quality time at the beginning of the project to draw out the structures, if these are not already established across the company by the project control department.

We also have seen how these generate interconnections with other processes. Above all, we have seen that if processes are not clear and well-established at the start, we instil 'pathogens' that will re-surface at some point in the future and that to remedy will cause considerable pain to all parties.

2. Flexibility and adaptability of the structures

Ok, this is where I can hear you say *'projects change during their life-cycle; how do you want me to fix something from the pre-concept or pre-feasibility period*?

To note, no project management theory, nor this text, suggests the following:

1. That you set up the whole detailed structure to the lowest level at the initial stages of the project, unless there is a company-wide standard where the company delivers similar projects continuously, or
2. The structures are not / cannot be flexible.

Structures should allow for flexibility, even those for companies that have a standard type, or types, of project(s).

This can be done in various ways including:

a. Starting with a high level structure and expanding as the project progresses through the different stages. This can be seen as a kind of 'rolling wave' deployment of the structure. You should, however, have the delivery of the full project, in the case of the WBS, described in the structure.

b. Allowing for and including, when required, an interdisciplinary approach. By that I mean you allow for elements of work that are usually covered in one WBS element, for example construction, to be covered or done under a different WBS element, e.g. design. To expand on the example, in many occasions you have 'construction' works or 'installation' required during the 'design' period. In this case, nothing should stop you having 'enabling works' as part of the design period / WBS element.

c. Including redundancy. This means that when you are developing the structure allow for the unexpected. Examples of this flexibility:
    1) In a client project where you are involved at the early stages, e.g. feasibility and the concept stage is to be completed by the Asset Management department, allow for a pre-feasibility work level;

    2) In a contractor construction project, allow for a 'pre-detailed' design level, or for a level which indicates works prior to starting the main ones (perhaps with your enabling works level);

    3) At a low level of the WBS, under the specific work packages, it would be beneficial to consider allowing for a 'change' capturing part. This would

allow you to capture all changes that are approved by the project manager under the package of work. Let's not forget that many of us use the WBS to produce reports and this will give you some flexibility in reporting approved changes, or even pending ones.

The structures should be able to incorporate flexibility – especially in the description (of the structure elements / nodes). It is lack of early thinking that generates the restrictions which stop us responding to a change or an additional requirement. Even in the case where we have set up the 'coding' and we have nicely indicated the drops in levels as, 1.01, 1.02, etc. we can still include 'dummy / empty' codes and allow for a new element to be introduced between the two numbers / codes.

We have to be able to adapt to change since it is impossible to capture everything correctly from the beginning. Nothing in the structures themselves should stop you from having the required flexibility.

There are also other project management processes that would help us on this.

Before a project starts the team holds a Value Management (pre-construction) or even a Value Engineering workshop (during construction) with all the stakeholders present. Nothing stops us starting with the structures – WBS, OBS, RAM or any others.

Another process to consider is that of Risk Management with the required workshop. Again, and we will cover the Risk Breakdown Structure (RBS) below, start the risk workshop with a discussion on the WBS, OBS and the RAM. As you go through the levels and you dive into the lower ones looking for risks / opportunities and establish the responsible person / entity, ask the audience if something has been missed out from the higher level structures.

To note, we are focusing on flexibility in adding parts to the structures, but nothing should stop us asking if we need to remove parts.

However, there is a point where you should stop making changes to the structures. This point is when you / the project enters the relevant stage of works, typically the Design stage, or Construction / Implementation stage. By this stage, if you have forgotten something at a high level then something has likely gone drastically wrong earlier and I am not talking about project scope change.

One point that needs to be remembered is that when you make changes to the structure, and especially the WBS, you have to update the Project Control Plan (PCP) (or Handbook) as well as the Project Management Plan (PMP). Both documents are some of the main communication tools to the project team members.

So:

| | |
|---|---|
| Flexibility? | Yes. |
| Thinking ahead? | Yes. |
| Accepting change? | Yes. |
| Using other processes to support the thinking about structures? | Yes |

But do not forget to communicate and update the standard project documentation, as you do this.

3. Structures and managing the effects of complexity

We have to accept that the theory of project structures has been around for few decades and we should also accept the truth that this is not adhered to; I suppose that is one of the reasons for my decision to write this book. However, in the last few years other theories have emerged which start to consider the reasons for this. One of these theories is complexity. And by that I do not mean the technical one that we know exists in projects, but the more generic one, the one that deals with the dynamic interconnections between 'systems'.

Although, there is no one specific definition of complexity, the majority agree that complexity is '*the natural consequence of the interaction of dynamic systems and relationships*'. So it is the theory that deals with the interconnections between the systems and not the systems themselves. In complex systems, such as the projects we are dealing with and delivering, interactions reinforce one another and result in behaviour that is very different from the norm (Miller & Page, 2007:50).

In my PhD thesis (Antoniadis, 2009) I considered that if we know the characteristics of complexity and their effects, as we do in all sciences, then we should be able to manage its effects. Therefore, considering the project management processes and how are these affected by the complexity characteristics, we should be able to identify actions that we can take to manage these effects.

There is a specific complexity characteristic that affects project management structures, or more specifically the use of the structures as a tool to manage projects. This complexity characteristic is that of 'downward causation' (Lucas, 2000b). 'Downward causation' is used to describe when '*the existence and properties of the parts themselves are affected by the emergent properties (or higher level systemic features) of the whole*'. In the case of the project structures, and for those parties involved in the project, the systemic features of the whole WBS affect the individual parties that are brought together (emergence) to deliver the project. In other words, the structures form a 'straight jacket' for the parties delivering the projects. We will not expand any further on this as it means we are moving to a different subject (which is clearly interconnected). However, we can see that structures allow the project team to manage the effects of complexity.

4. The project management structures

The primary and most common standard structures are the WBS and the OBS. Every project textbook and the APM / PMI BoKs, refer to these structures. After a brief description, we will look at some examples.

Why these, I hear you ask?

Well, if you look at what they describe the answer will be obvious. It is about the work packages that deliver the project (WBS) and the teams that will deliver the project (OBS).

There are also supporting structures. These are required because they either link information between different processes or are used to generate lateral reports when building programmes. These reports cut horizontally, and not vertically, our information datasets. For example, reports that cut across the structures usually draw from the CBS, or Document Control Structure, or Risk Breakdown Structure (RBS), or Programme / Portfolio Breakdown Structures.

In addition to the above we also have, as mentioned earlier, a structure that is generated from the combination of the two primary ones. One of the most known structure combination outputs and a common project management one is the RAM.

We will look now at the structures and I will give a few examples.

1. Work Breakdown Structure (WBS)

The WBS is the hierarchical breakdown of project scope that is delivered through the relevant work packages.

But let's remind ourselves the basics of the WBS.

| **WORK** | **BREAKDOWN** | **STRUCTURE** |
|---|---|---|
| What you have to do.<br>Your work packages that will become your:<br>➢ Tasks.<br>➢ Activities | ➢ Broken down progressively to lower levels of detail.<br>➢ Below the lowest WBS level you have the programme activities.<br>➢ It is well defined. | ➢ logical organisation.<br>➢ tasks grouped appropriately. |
| **WBS is NOT about:**<br>➢ The order of how things are done, or<br>➢ How long it will take, or<br>➢ Who will carry out the tasks, or even<br>➢ The number of people / teams / material, etc. | | |

The WBS has to be considered as the basis around which everything else can and has to be built. But let's look at some examples, first of the style and from real life.

*Figure 3.1:* Example 1 of Work Breakdown Structure (WBS)

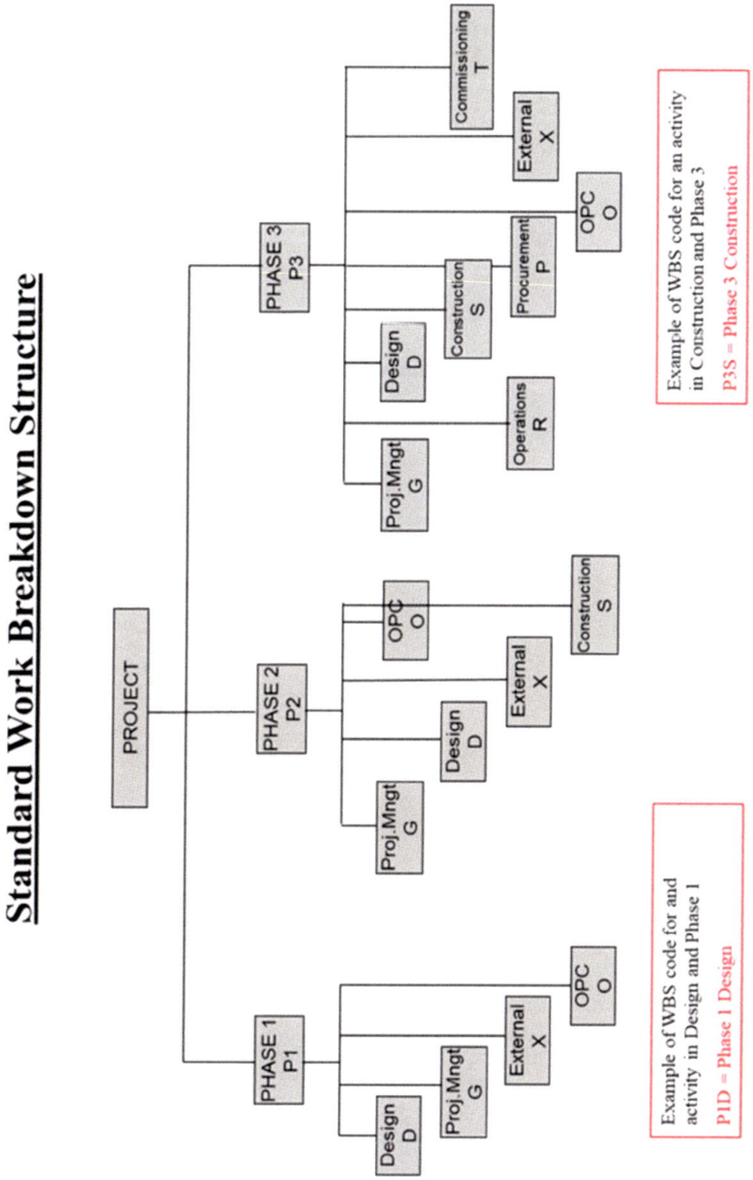

Figure 3.2: *Example 2 of Work Breakdown Structure* (WBS)

| Level 0 | Level 1 | Level 2 | Level 3 | Level 4 |
|---|---|---|---|---|
| Project | | | | |
| | Management | | | |
| | Detail Design | | | |
| | | Civil | | |
| | | M&E | | |
| | | Structural | | |
| | Site Logistics | | | |
| | | Site Set up | | |
| | Construction | | | |
| | | Site Preparation | | |
| | | Ground Works & Basement | | |
| | | | Block A | |
| | | | | Columns & Walls |
| | | | | Basement works |
| | | | Block B | |
| | | | | Columns & Walls |
| | | | | Basement works |
| | | | Block C | |
| | | | | Columns & Walls |
| | | | | Basement works |
| | | | Block D & E | |
| | | | | Columns & Walls |
| | | | | Basement works |
| | | Block A | | |
| | | | Ground floor & Substructure | |
| | | | Superstructure | |
| | | | Cladding & Roofing | |
| | | | Services - M&E | |
| | | | Fitout | |
| | | | Commissioning | |
| | | Block B | | |
| | | | Ground floor & Substructure | |
| | | | Superstructure | |
| | | | Cladding & Roofing | |
| | | | Services - M&E | |
| | | | Fitout | |
| | | | Commissioning | |
| | | Block C | | |
| | | | Ground floor & Substructure | |
| | | | Superstructure | |
| | | | Cladding & Roofing | |
| | | | Services - M&E | |
| | | | Fitout | |
| | | | Commissioning | |
| | | Block D | | |
| | | | Ground floor & Substructure | |
| | | | Superstructure | |
| | | | Cladding & Roofing | |
| | | | Services - M&E | |
| | | | Fitout | |
| | | | Commissioning | |
| | | Block E | | |
| | | | Ground floor & Substructure | |
| | | | Superstructure | |
| | | | Cladding & Roofing | |
| | | | Services - M&E | |
| | | | Fitout | |
| | | | Commissioning | |

Figure 3.3: *Example 3 of Work Breakdown Structure* (WBS)

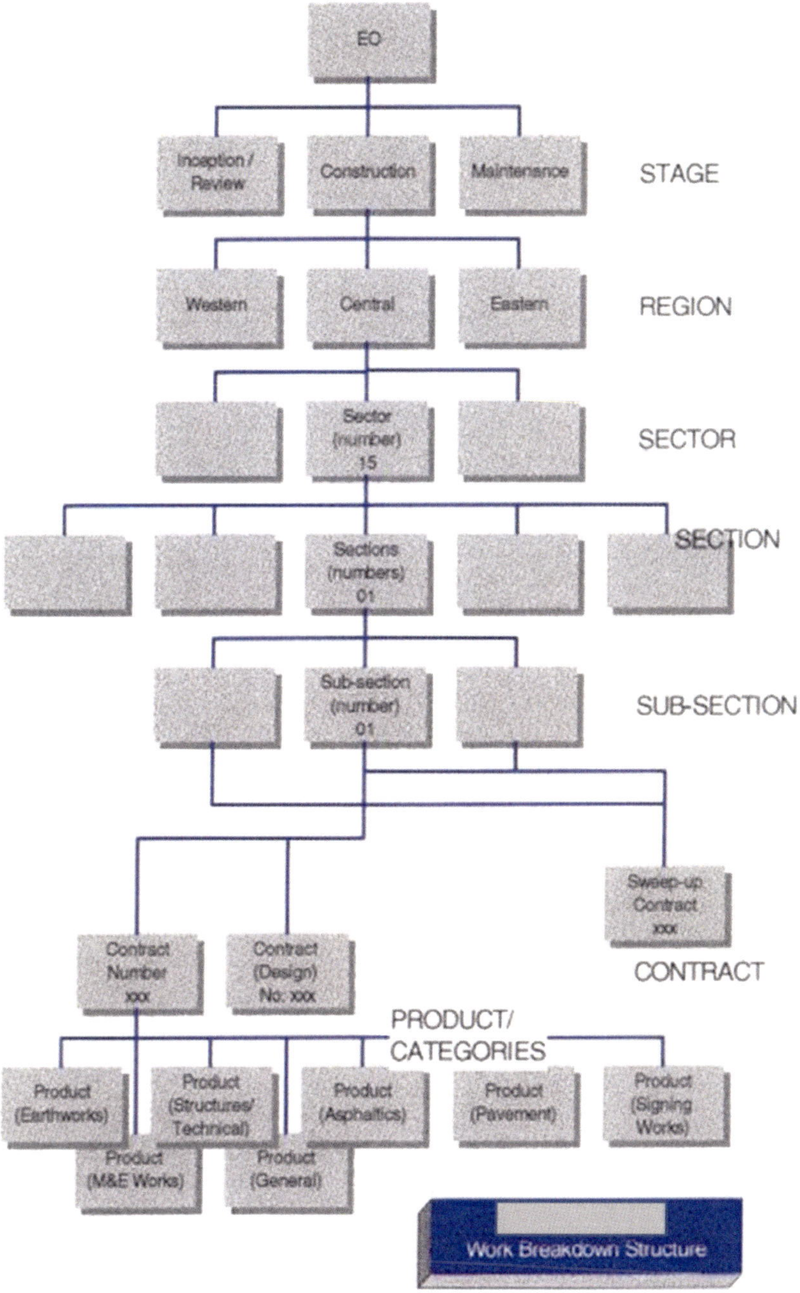

2. Cost Breakdown Structure (CBS)

The APM (2015) describes the CBS as the structure that '*defines the level at which costs will be collected*' and indicates that it does not necessarily have a hierarchy as the WBS does. However, from a practical perspective I would say in all the environments I've experienced we have always created a hierarchical CBS.

By generating a structure, and mainly for project control of client organisations, you are able to set up not only the basis for capturing costs but also to prepare initial project estimates which will allow you the evaluation of submitted bids. You are also able to gather information and roll this up to other relevant processes.

In the two examples, see Figures 3.5 & 3.6, you can see typical project CBS. Additionally, Figure 3.6 shows how you can use the CBS to establish the project cost reporting mechanism, where the use of cost coding allows the user to report against certain cost elements of the project.

Figure 3.4 demonstrates how, starting from the process level, you can break down the costs that need to be considered under projects that deliver the specific process.

In addition to the above, the importance of the CBS as a structure is highlighted in the commercial approach to partnering, where cost elements can be used from the tendering stage to the delivery stage by using the cost units for the Applications for Payment (AFP). An example, as evident in Figure 3.4, would be AFPs submitted at Level 4 where these are supported by the structure at the lower levels and can be interrogated at any point in time.

The CBS, as described earlier has a very important role to play in managing projects. As well as supporting the WBS, it is the main structure that links the sub-process of cost control with those teams on the finance side. Therefore, it is critical to establish this early on, if it does not already exist as a standard process from the company side.

Figure 3.4: *Example 1 of Cost Breakdown Structure* (CBS)

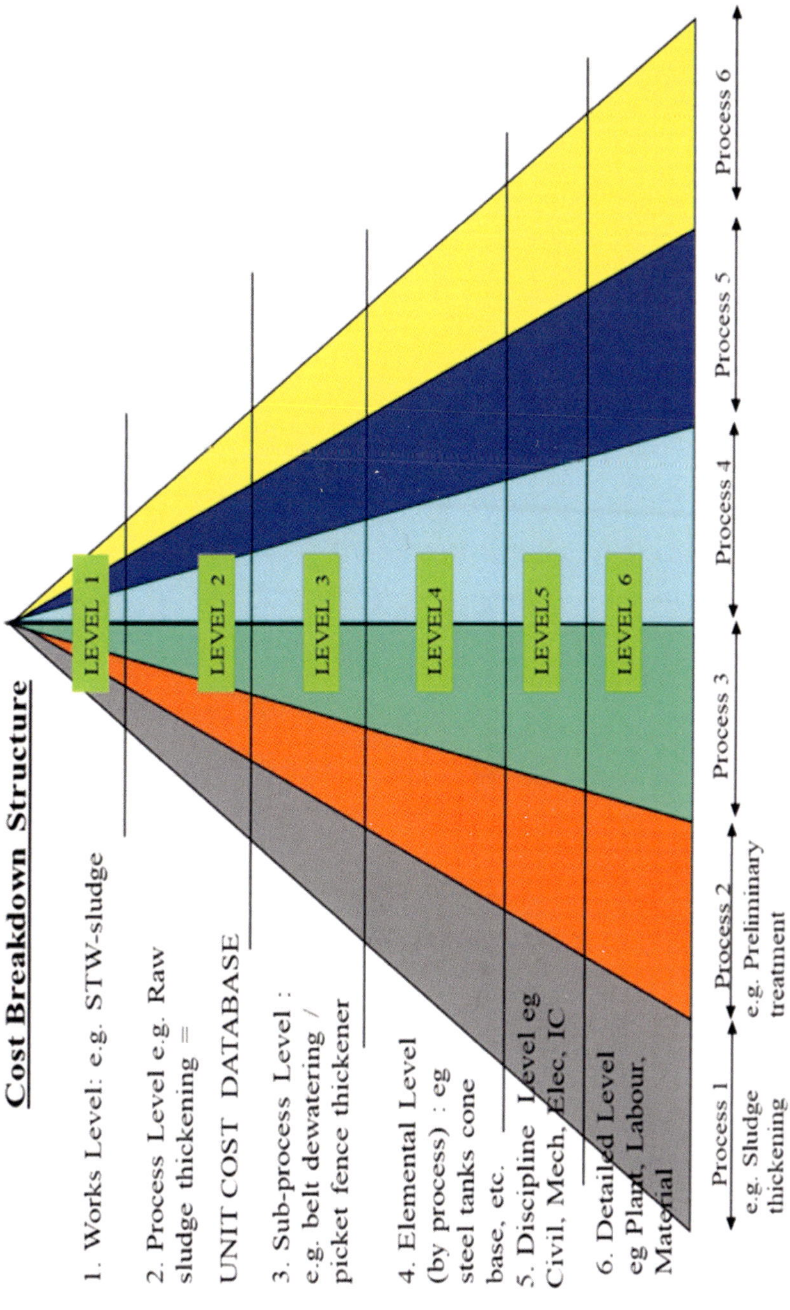

Figure 3.5: *Example 2 of Cost Breakdown Structure* (CBS)

Figure 3.6: Example 3 of Cost Breakdown Structure (CBS)

**Cost Proforma with CBS Coding**

| Cost Breakdown Structure & Cost Report | | | | | | | | |
|---|---|---|---|---|---|---|---|---|
| Report LEVEL | | 1 | 2 | 3 | | 1 | 2 | 3 |
| A.1.1.1 | A | 1 | 1 | 1 | 1.1.1 | Enabling Works | Enabling Works | Outstanding Structural Survey 1st Cut Authorised |
| A.1.1.2 | A | 1 | 1 | 2 | 1.1.2 | | | Outstanding Structural Survey (additional funds) Update |
| A.1.1.3 | A | 1 | 1 | 3 | 1.1.3 | | | People Moves £k 1st Cut Authorised |
| A.1.1.4 | A | 1 | 1 | 4 | 1.1.4 | | | People Moves £k (additional funds) update |
| A.1.1.5 | A | 1 | 1 | 5 | 1.1.5 | | | People Moves £k New Authorisation |
| B.1.1.6 | B | 1 | 1 | 6 | 1.1.6 | | | Equipment Recovery Contract Authorised |
| B.1.1.7 | B | 1 | 1 | 7 | 1.1.7 | | | Equipment Recovery Contract Additional Update |
| B.1.1.8 | B | 1 | 1 | 8 | 1.1.8 | | | Equipment Recovery Contract New Authorisation |
| B.1.1.9 | B | 1 | 1 | 9 | 1.1.9 | | | Equipment Recovery NB Authorised |
| B.1.1.10 | B | 1 | 1 | 10 | 1.1.10 | | | Equipment Recovery NB Additional Update |
| B.1.1.11 | B | 1 | 1 | 11 | 1.1.11 | | | Equipment Recovery NB New Authorisation |
| B.1.1.12 | B | 1 | 1 | 12 | 1.1.12 | | | DMSU Recovery Authorised |
| B.1.1.13 | B | 1 | 1 | 13 | 1.1.13 | | | DMSU Recovery Additional Update |
| B.1.1.14 | B | 1 | 1 | 14 | 1.1.14 | | | DMSU Recovery New Authorisation |
| C.1.2.1 | C | 1 | 2 | 1 | 1.2.1 | | Contractors Fee's | Contractors Fee (ECC form of contract) |
| C.1.3.1 | C | 1 | 3 | 1 | 1.3.1 | | Assessment of Contractors Enabling Works | Structural Engineer |
| C.1.3.2 | C | 1 | 3 | 2 | 1.3.2 | | | Electrical Engineer |
| C.1.3.3 | C | 1 | 3 | 3 | 1.3.3 | | | Report Production |
| C.1.3.4 | C | 1 | 3 | 4 | 1.3.4 | | | Production of enabling works sketch |
| C.1.3.5 | C | 1 | 3 | 5 | 1.3.5 | | | Site Manager |
| C.1.3.6 | C | 1 | 3 | 6 | 1.3.6 | | | QS |
| C.1.99 | C | 1 | 99 | | 1.99 | | | Star Rate |
| C.1.4.1 | C | 1 | 4 | 1 | 1.4.1 | | Asbestos | Asbestos Survey (level 2) |
| C.1.4.2 | C | 1 | 4 | 2 | 1.4.2 | | | Asbestos Survey (level 3) |
| D.1.4.3 | D | 1 | 4 | 3 | 1.4.3 | | | Asbestos removal/Encapsulation |
| D.1.5.1 | D | 1 | 5 | 1 | 1.5.1 | | Contractors Enabling Works | Structural Strenghtening |
| D.1.5.2 | D | 1 | 5 | 2 | 1.5.2 | | | Builders Work/Strip out/Relocate/M & E |

3. Organisational Breakdown Structure (OBS)
The OBS is defined as the hierarchical structure that describes the line management, the communication and escalation of issues as well as the reporting lines / structure within a project.

In the following figures I present some standard models as well as some examples from practice.

Figure 3.7: *Example 1 of Organisational Breakdown Structure (OBS)*

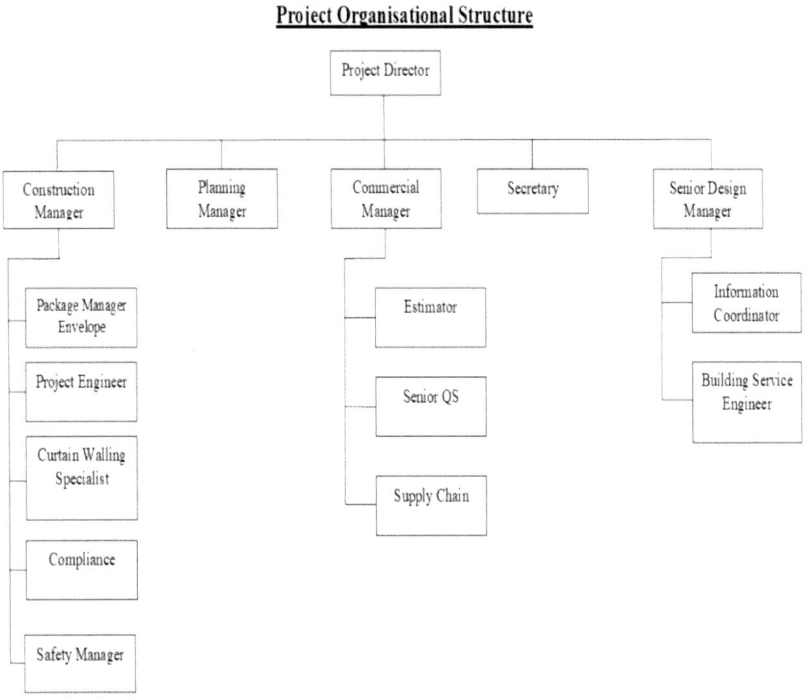

Figure 3.8: *Example 2 of Organisational Breakdown Structure* (OBS)

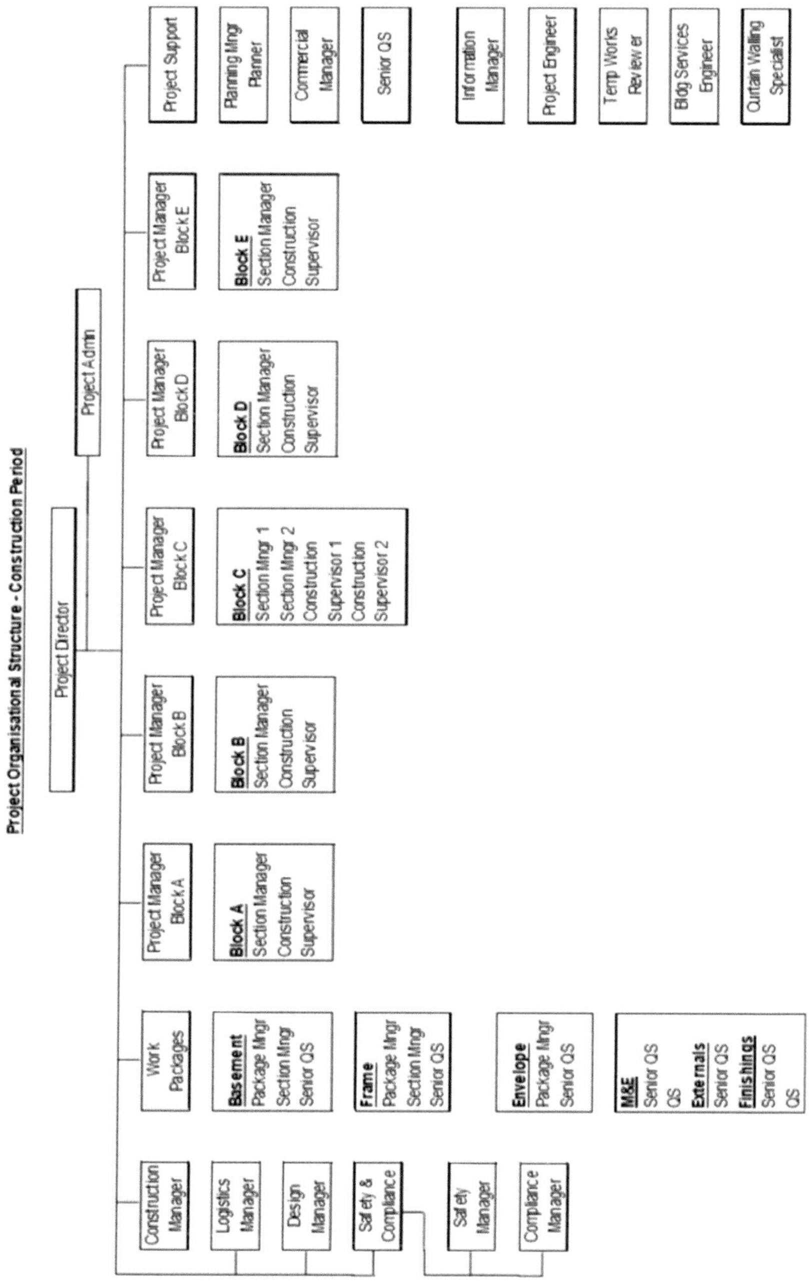

Figure 3.9: Example 3 of Organisational Breakdown Structure (OBS)

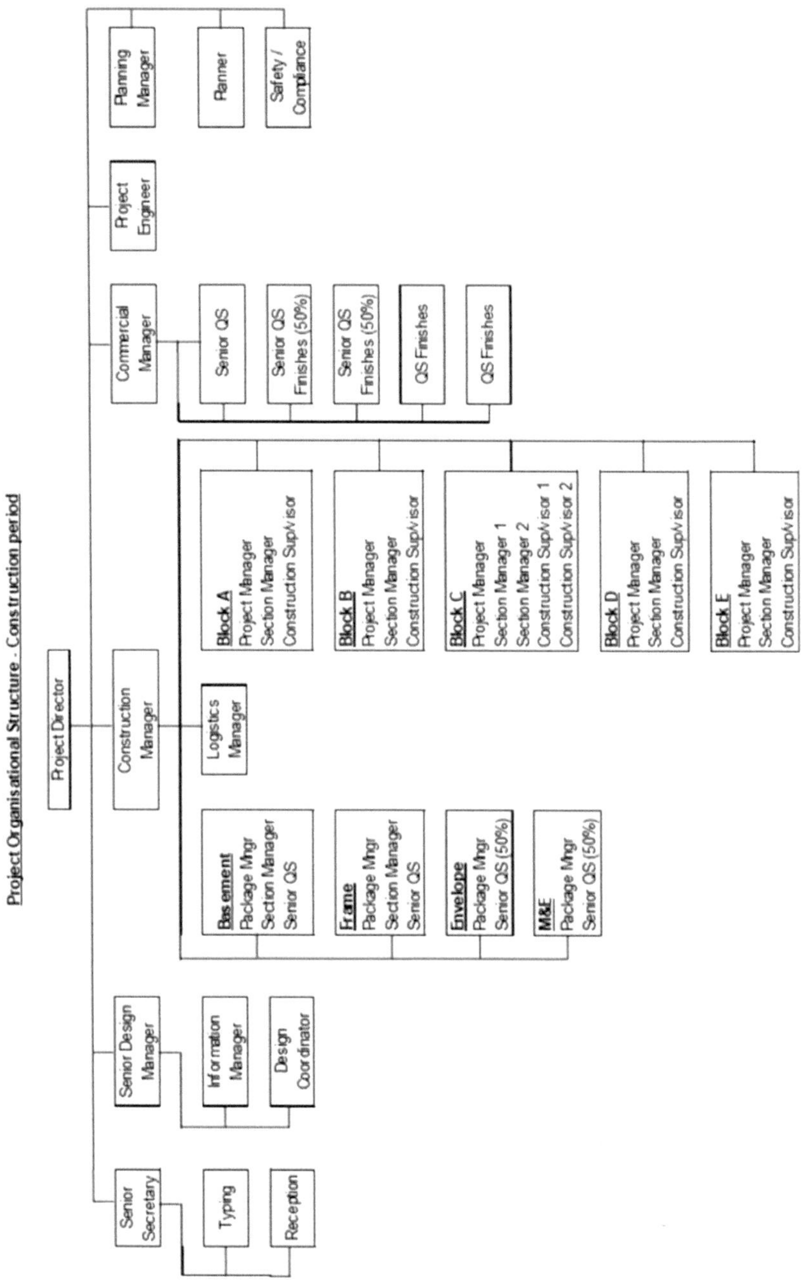

Figure 3.10: *Example 4 of Organisational Breakdown Structure* (OBS) – *at a wider project / programme level*

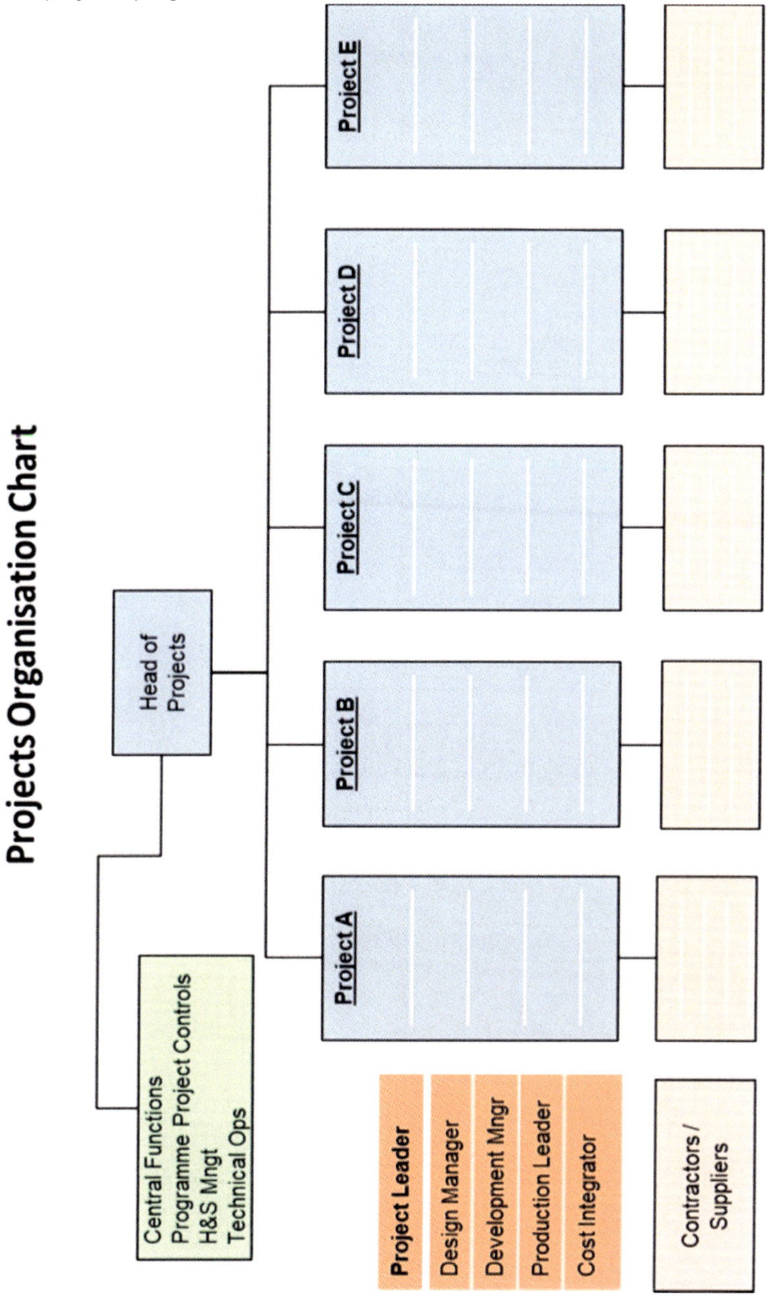

4. Programme Breakdown Structure (PgBS)

To my knowledge there is no definition applied yet for PgBS.

In practice, a number of organisations that deliver programmes of work set up a structure in order to roll up information and monitor and report to Senior Management levels. Usually information is rolled up to the projects that are delivering the parts of the programme and these then are given a RAG (Red, Amber, Green) status.

I have given below an example (Figure 3.11) from a £10Bn worth programme of works that covered the whole of UK where we had to roll-up information from projects to 'chains', to geographical regions and then for the overall programme.

In this particular case we called the structure a PgBS. However, if you look back at Figure 3.3 on a similarly mega project overseas, in that case it was decided that the structure will be called WBS.

Figure 3.11: *Example of Programme Breakdown Structure* (PgBS)

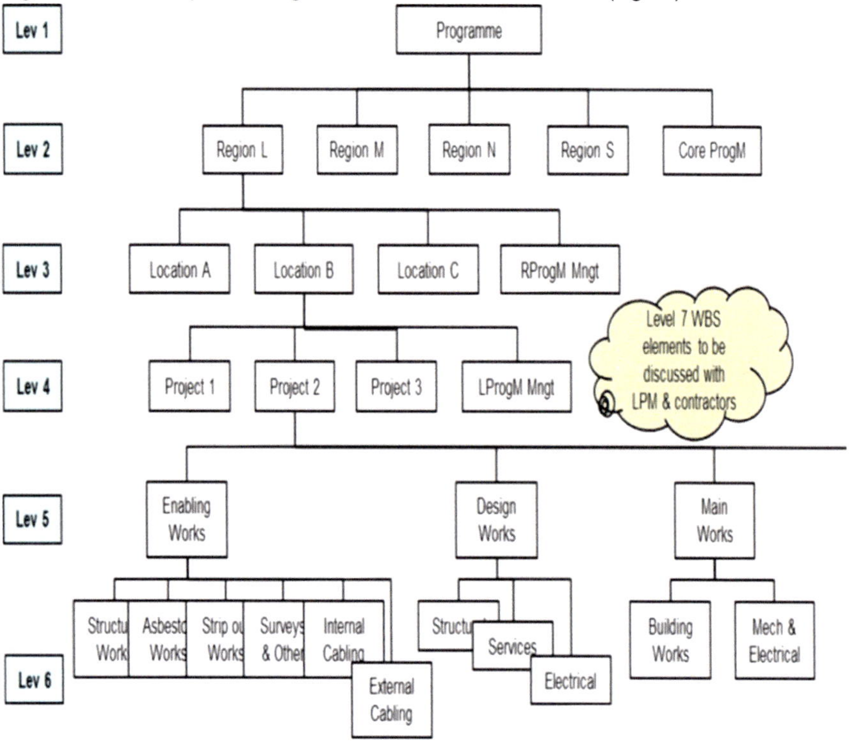


5. Project / Product Breakdown Structure (PBS)

The APM (2015) describes PBS as the structure that '*provides the relationship between the products and sub-products such assemblies, sub-assemblies and parts that contribute to the delivery of the project scope*'. It is a hierarchical structure of the products and outcomes delivered by the project. It should not include work packages as such because if it does then it has been confused with the WBS. However, the PBS is used for the derivation of the WBS and also forms the base from which the execution strategy can be derived and for the configuration management process.

Figure 3.12: *Example of Project / Product Breakdown Structure* (PBS)

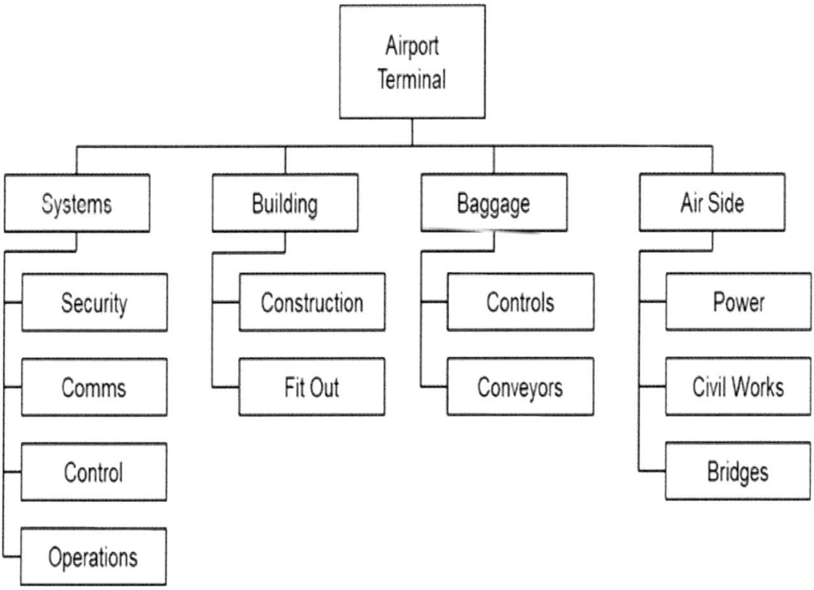


6. Risk Breakdown Structure (RBS)

Before we discuss the RBS it has to be made clear that when we talk about risks we include opportunities as well and that this structure reflects both of these.

Although we consider structures as part of projects, in many occasions these support, much wider requirements. The RBS is one of these structures that can be rolled down from company-wide level to project level. An example of this for an Asset Management organisation is shown in Figure 3.13b. The asset hierarchy reflects relevant risks down to a certain level, and therefore the relevant coding. Below that the project risks are broken down into the various categories.

Figure 3.13a represents an example of a company-wide RBS where, in addition to Directorate risks, we can have Programme of Asset level threats or opportunities that these roll-up or cascade down the structure.

Figure 3.13a: *Generic example of Risk Breakdown Structure* (RBS)

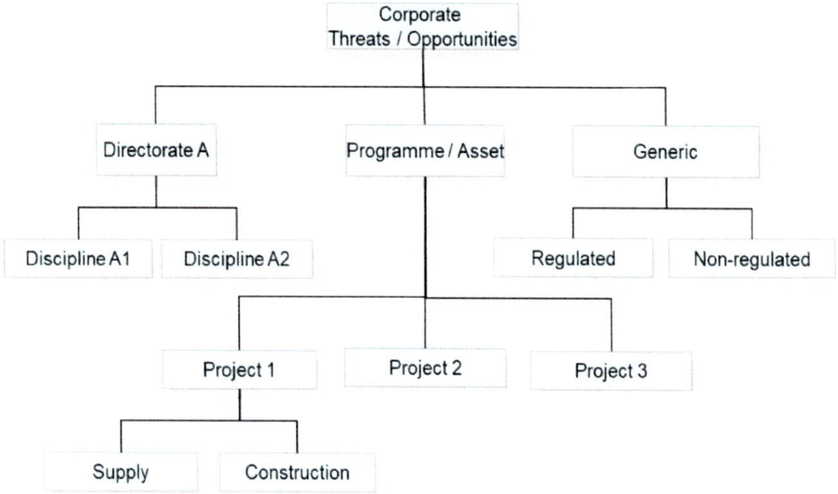


Figure 3.13b below presents an example of the RBS from a utilities organisation.

Figure 3.13b: Utilities organisation example of *Risk Breakdown Structure* (RBS)

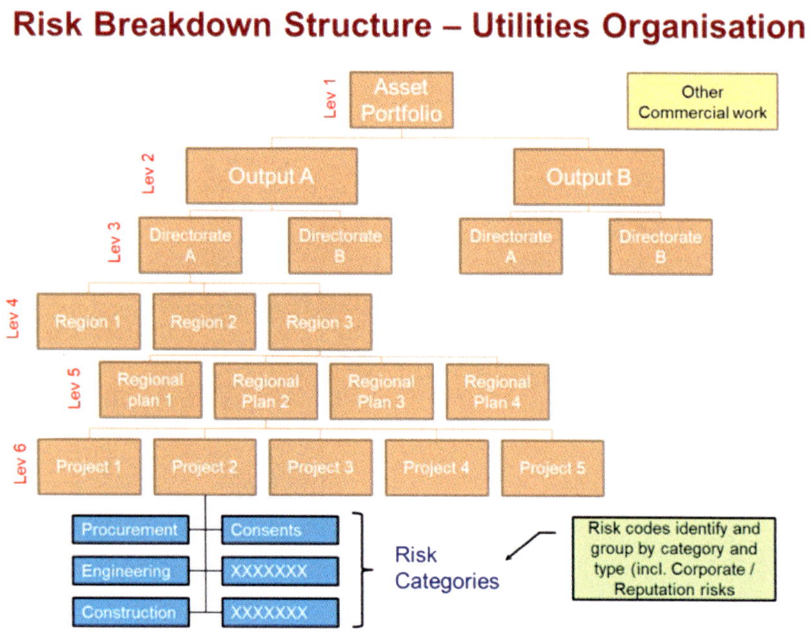


Grouping risks by hierarchical category enables ease of reporting by allowing for the rolling up of risk exposure levels, addressing company-wide risks with a more strategic point of view, e.g. procurement / contractual arrangements, and a number of other advantages.

By assigning categories, risks could be grouped by various topics. For example, corporate risk, contract (especially if these are shared with Partners), external organisations, etc. Categories could also be used to group risks that might span different Directorates or departments into 'parent' and 'child' as in Figure 3.13b.

Another advantage of the RBS is that, at risk identification stage and whilst the team is looking at the WSB packages, it can be used during risk workshops to consider areas / topics that could be possible sources of risks or opportunities.

A point to note here is that if the RBS is not set up centrally you (PM or project controller) will have to set this up from scratch. So, how do you go about doing it?

Using the risk workshop to identify potential project risks, consider grouping the output by category. Categories could be used as an umbrella for lower level risks so that they can be rolled up for reporting purposes. These could include:

- Design;
- Consents;
- Procurement;
- Contractual;
- Construction and Commissioning.

It has also been suggested that you could start the risk management workshop by considering risks under the respective WBS elements. This will allow you not only to categorise the risks but also link them to the WBS.

Generating the RBS, the relevant categories and the risk registers at the beginning of the project not only allows for their management during the project life cycle, but also minimises the 'pathogens' introduced at the early stages of the project as much as possible. These pathogens will incubate in the project until the prevailing conditions spark their revival and create another issue for the team. Early engagement, understanding and consideration of risks and their structure can better prepare the team for the future. As these risks are then 'managed' or 'impact', the project register can be updated. The relevant detail of how this can be done is covered in basic risk management theory.

Whilst highlighting the importance of creating risk categories it was mentioned that this is beneficial for partnering contractual arrangements. The reason is that as risks are defined, they can be grouped by the RBS coding and categorised and therefore easily reported at various levels. Clarity resolves future challenges and monitoring of projects becomes less confrontational. Of course this does not mean that 'interesting' issues and discussions will not be necessary later on in the process, but it allows, at early stage the identification and handling of pathogens. Also, and this is of comfort to the Commercial / Procurement colleagues, structures work to their benefit as well. By setting up the standard project structures at an early stage it is possible to manage the effects of complexity characteristics which impact the risk management process.

Figure 3.14 presents another practical example of the use of RBS in a project concerning the handover of IT rooms.

Figure 3.14: *Example of Risk Breakdown Structure* (RBS)

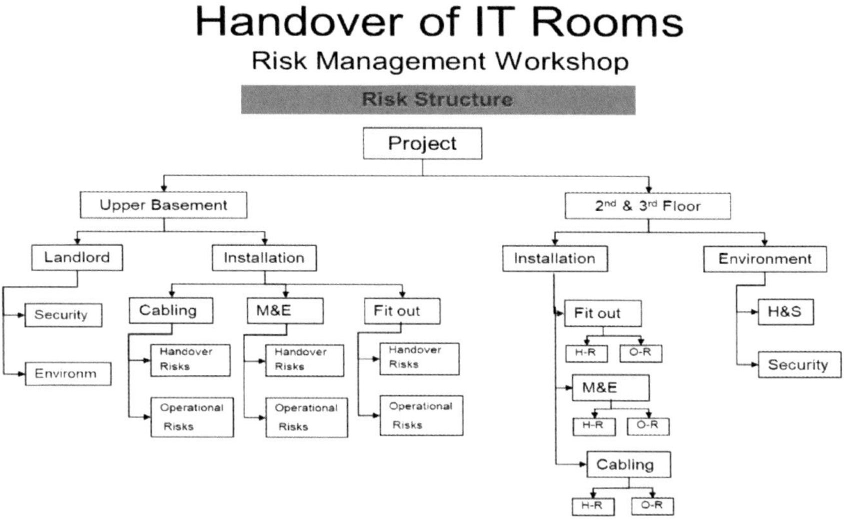


7. Document Breakdown Structure (DBS)

Currently there is no standard definition given for structuring project documentation. However, every project and every organisation has (or should have) a Document Control system that is used for various reasons – legal, financial, design, etc.

*Can we really class the DBS as one of the standard project structures?*
Well, let us for a minute consider what is one of the main means of communication in project.

Document control systems, usually based on software packages, require the classification of documents by type, in-coming or out-going, etc. Standardising the project documentation using a structure is an obvious advantage, even if there is not a central direction given. The structuring of the coding enables ease of storage and retrieval, monitoring of actions required and completed, which are just some of its advantages.

We will not go into the detail of what is considered as project documentation here, but must accept that for every project there will be plenty of it. Also, in addition to the standard, by now drawing numbering requirements, we have to be able to record every document on the project.

Let's take the simple example of the project (or any other level) monthly reports. These, in addition to the usual function of reporting status, for various reasons, have to be stored for a certain number of years. Without the DBS it will be extremely difficult to search through the archives in order to retrieve the correct document.

Obviously the DBS has to be dependent on the WBS. Standardising the documentation requirements in a structured manner enables:

- Appropriate organisation;
- Ease of retrieval;
- Consistency;
- Monitoring, (by simply identifying level of action(s) required you can produce a quick end of the week report on 'outstanding actions and urgency').

Also, as mentioned briefly above, the project DBS should link to company-wide systems like Document Management, Correspondence systems, Archives, Drawing register etc.

Figure 3.15 below gives an example of a project DBS. In this particular case the coding given to each document, before the levels shown, were the WBS, CBS, Product and Scope. The other structures were included in the coding in order to 'position' the document in the appropriate document folder.

Figure 3.15: *Example of Document Breakdown Structure* (DBS)

| Document Management | | | |
|---|---|---|---|
| Type of Document | In or Outgoing | Revisions | Version |
| (xxxx) | 'I' or 'O' | (xxx) | (xxx) |
| G01-G99: General document | | | |
| CD01-99: Cadastral document | | | |
| R01-99: Report | | | |
| A01-99: Tender document | | | |
| B01-99: Contract document | | | |
| J01-99: Invoice | | | |
| T01-99: Time/Schedule | | | |
| QA01-99: QA/QC report | | | |
| C01-99: Cost report | | | |
| L01-99: Correspondence | | | |
| S01-99: Standards | | | |
| Z01:99: Digital submission | | | |

5. Combining project management structures

1. Results from combining the structures

As was highlighted earlier we can produce a very important output when two primary structures are combined.

Combining structures enables the project team to describe and clarify the interfaces between the different reporting parts of the project – work, cost, risk, etc. and therefore enable management of complexity arising from interfaces between the processes and the teams that operate using these.

You will see below some examples of the output from this combination of structures.

2. What do you get from WBS + OBS = RAM

The combination of the WBS and the OBS defines the Responsibility Assignment Matrix (RAM). It defines which part of the work or package(s) is delivered by which responsible discipline manager / team leader. This is either identified as a particular named person or a generic team leader e.g. Design lead, Construction lead, etc.

Figure 3.16 below represents a theoretical RAM and how it is brought together by combining the WBS with the OBS. The junction of the two structures leads to the RAM.

Figure 3.16: *Example of the RAM built by the WBS and the OBS*

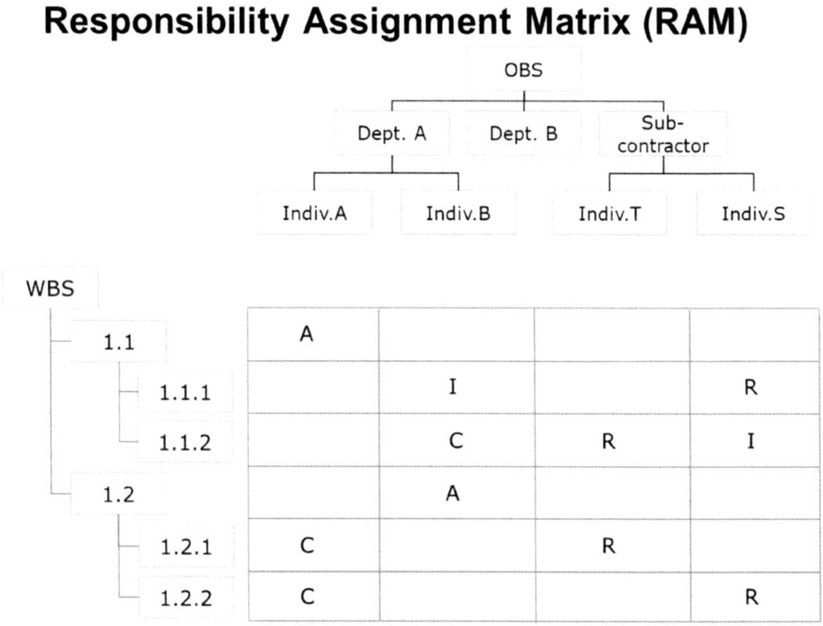


| WBS | Indiv.A | Indiv.B | Indiv.T | Indiv.S |
|---|---|---|---|---|
| 1.1 | A | | | |
| 1.1.1 | | I | | R |
| 1.1.2 | | C | R | I |
| 1.2 | | A | | |
| 1.2.1 | C | | R | |
| 1.2.2 | C | | | R |

The practical output from the RAM is the identification of a named individual who will have the responsibility to deliver the work package or part of work. Often these individuals are also the resource managers that are expected to report against the Labour, Plant, Material, Subcontractor costs / resources of that package.

This link gives us an initial indication of how two primary structures can connect. We will see below how this links to the CBS, both on cost and resource (as in person hours, number of plant, cost of material, etc.).

When used properly, the RAM can contribute to a number of other benefits from both the line management and project management sides.

From the line management side, the individual identified under line management, this person is responsible for team members' well-being, quality of the output, implementation of departmental procedures, career development, early engagement with the 'mother—ship', etc. For example, releasing individuals back to the department / organisation, ordering of plant, material and other work package requirements, and HR issues.

From the project management side, the person identified in the relevant RAM box is responsible for monitoring and discussing progress, resolving appropriate level project issues, coordinating the relevant output, reporting to the PM. That person also raises issues to the supporting department or organisation with regard to supplies, resource, etc. At project level the RAM also allows for the allocation of owners to risk mitigating actions. By simply identifying the risks under the respective work packages the RAM indicates who the primary owner of the mitigating action is.
The examples given above are some of the activities performed by the person identified. Depending on the organisation the 'Responsibilities' will vary, but the important point is to set these duties up.

3. Combining other structures

Generally, the only output mentioned in books and BoKs from combining two structures is the RAM. However, in practice we can bring together other structures that will support project control and produce different outputs.
We will discuss below some examples.

One such combination is that of WBS-OBS-CBS. Figure 3.17 below gives you an indication of the outputs from this, which, apart from the RAM generated by the WBS and OBS, delivers a number of other project control outputs. For example, from the programme you can extract the tasks and the job cards, resource histograms, project S-curves and provide cost information upstream to Finance for the monthly accounts.

Figure 3.17: *Example of three Structures* (WBS-OBS-CBS)

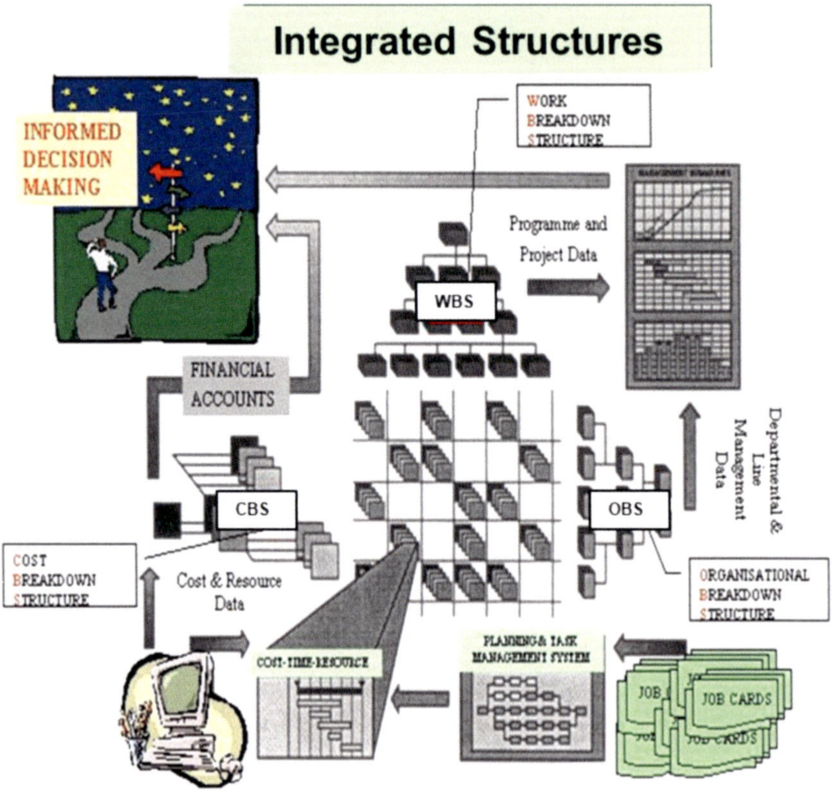


Let us look at another pairing, that of WBS and CBS.

In earlier chapters, it was noted that the general perception is that WBS and CBS cannot be brought together. However, if you think about it, the WBS can go down to as low a level as the company or the PM require. Then, below the lowest WBS level are the activities, logically linked and cost loaded, which describe how the team plans to deliver the work. If we consider the 'cost' / 'resource' side of the activities, these are made of the Labour, Plant, Material that are required to deliver the project / work package.

So, at that level we have the lowest, or perhaps the penultimate, level of the CBS coming into action. If then we follow the CBS upwards we can see how the cost can be rolled up according to the company cost centres, processes, external contributors, etc.

The examples below show how this has been implemented in two different utilities companies.

Figure 3.18: *Example 1 of combining WBS-CBS Structures*

| Programme Level WBS - CBS Structure ingtegration | | | |
|---|---|---|---|
| **WBS Code** | | **Level 1 CBS Codes & Descriptions** | |
| EW030 | Contractros | C | Assessment of Contractor Enabling works |
| EW050 - 080 | | D | Contractors Enabling Works |
| MT990 | | E | Contractors Fees |
| MT900 | | F | Design |
| MW100 | | G | Builders Works |
| MW300 | | H | Mechanical & Electriccal Works |
| MW500 | | J | Specialist Plant |
| MW300 | | L | AC Major Infrastructure |
| MW300 | | T | Contractor Cabling |
| MW300 | | U | Contractor Security |
| EW010 | Programme Management, Location Programme Management, Client | A | People move |
| EW020 | | B | Equipment Recoveries |
| EW030 | | K | Power Upgrade |
| A990 | | M | Technical Advisor |
| LZZ,MZZ | | N | Location Programme Management |
| NZZ, SZZ | | | |
| A990 | | P | Programme Management Team |
| A990 | | Q | Contingencies |
| A990 | | R | Planning |
| A990 | | S | Legal |
| EW040 | | T | Cabling |
| A990 | | U | Security |
| A990 | | V | Resilience |
| A990 | | W | Interim Funding |

Figure 3.19: *Example 2 of combining WBS - CBS Structures*

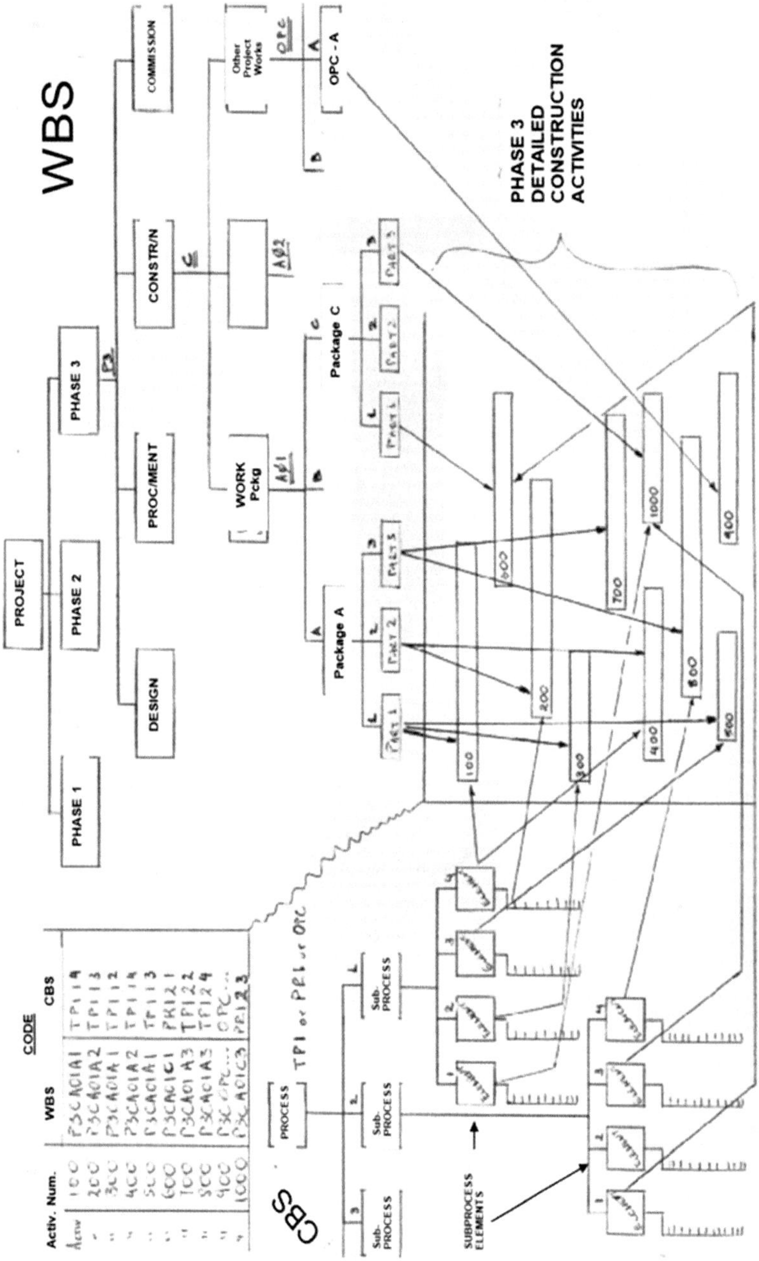

6. Dealing with other project management processes

*Can we have Risk as part of the WBS?*

Although process-wise it is not appropriate to remove the risk element from the respective WBS element(s), depending on the requirements, you can assign a WBS element that will capture all 'unallocated' risk monies.
Of course this means, as in the case of Heathrow Terminal 5 for example, the Client organisation has accepted and carries all the project risks (as a contingency) right from the initial stages of the project. At the early project stages, when not much has been decided about delivery, this approach allows for the project control to capture management requirements and present project contingency in the schedule.

Therefore, as the project progresses you have the means to distribute risk allowance to the appropriate work package / WBS element as and when packages are identified and risks can be allocated. At the same time you can report upwards the level of the unallocated risks, allow for project funding strategic decision making, reallocating funds and other relevant actions.

This approach could also be combined with the estimating process, considering that at the very early project stages estimates (depending on the method used) carry considerable allowances in addition to contingencies.

The example below, Figure 3.20, is one of a client environment. However, this does not mean that this approach cannot be used by contractor organisations. The approach taken in that particularly large project was to identify unallocated sums / contingency per package of work, in addition to the monies identified as risks and opportunities. Therefore, as orders were issued and works per package commenced we monitored changes and withdrew the monies from the unallocated funds. This was monitored monthly in the project programme and taken into account when reports were issued.

It can be used as a method when the brief is not clear and where in discussions with the client an informed decision needs to be made based on the level of risk that can be accepted. This allows for transparency at future points and again the 'pathogens' introduced at the initial stages are considered. So, as the 'incubation' period expires, in this case there is a clear scope, discussions on risk reduction are held and the risks/threats become opportunities / monies released.

What needs to be made clear here, for the avoidance of any doubt, is that you cannot have the specific work package risks and opportunities under the 'generic' risk WBS element and at the same time duplicate these under the respective work package elements.

Figure 3.20: *Example of using contingency as a preamble to risks*

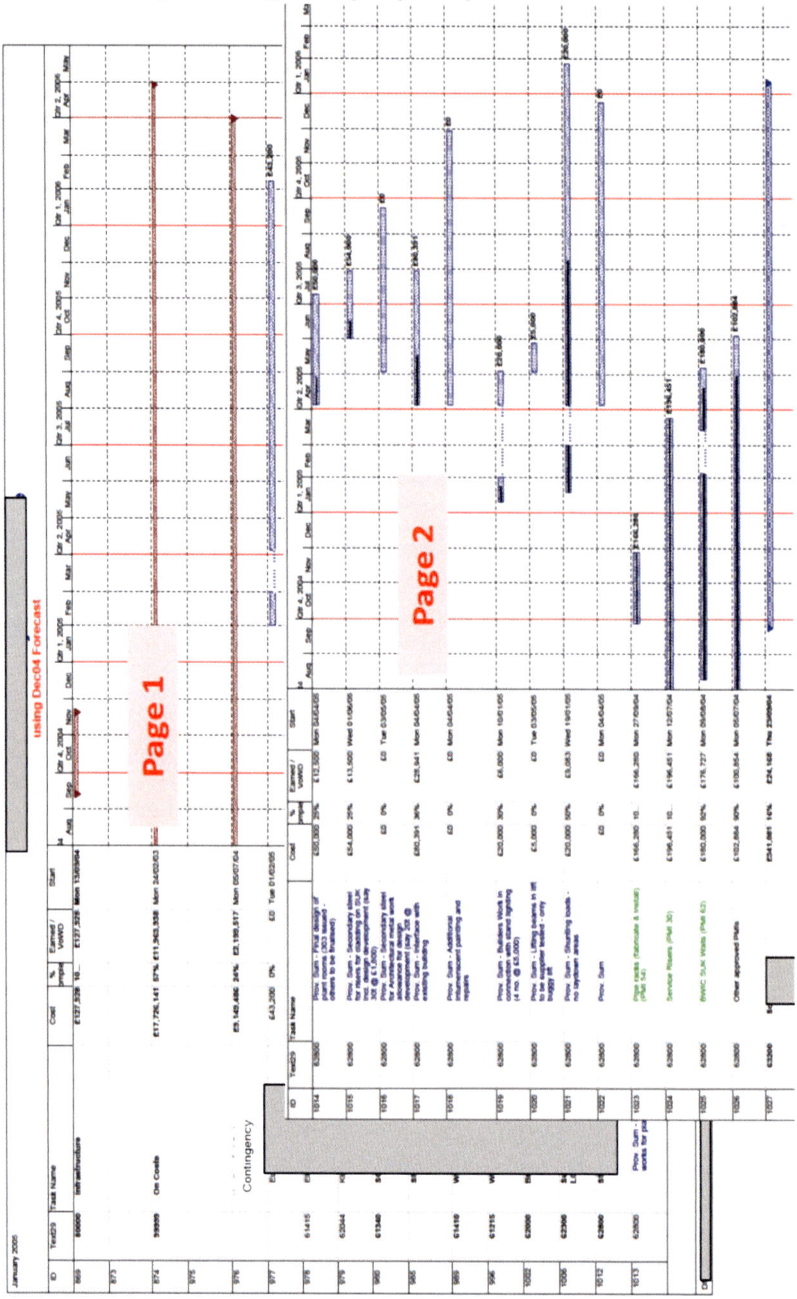

Another process that can be completed by combining the standard project management structures is that of Issues Management. Using the WBS, issues can always be referred to the appropriate WBS element and coded as such. Planning software tools have developed this capability and issues can be raised from within the tools and emailed across to the person / management level that is required to take the action(s). Issues can even be attached to Programme level structures and directed again towards the appropriate person.

As we have seen in this chapter the power of the standard structures is such that allows the project team, through project control, to improve the performance of the team and contribute towards the project management's success.

In the next chapter we will look into the issue of project control as a department, or perhaps something else that can be performed irrespective of specific roles.

# 4. IS PROJECT CONTROL A DEPARTMENT OR SOMETHING ELSE?

In Chapter 2 we covered part of the theory and the question 'what is project control'? In Chapter 3 we identified the project management theory structures that support the function of project control. So, now we need to consider the question of whether project control is a department or something else. By something else, I mean a set of processes performed by separate disciplines, or a part of another department or disciplines.

As we will see, this variableness is unique to project control and to the discipline compared to other disciplines within the 'control' or the project management environment.

To put some 'meat to the bone' on this issue, I will first describe the uniqueness of the issue. Then I will describe the various scenarios that should be considered by those that have the task of answering the question.

For the sake of simplicity when I refer to project I will mean a project or company. So, we have projects and project control performed by a number of different disciplines. Project control activities, depending on the organisation are performed by:

- The project planner;
- The quantity surveyor (but not including the planning side);
- The Project Manager;
- The person that extracts information from systems and then compiles a report (you can call this the project control report);
- The finance person tasked to compile 'the project control's report.

The confusion increases also from another side, that of job descriptions. In many occasions managers / companies give individuals the job title of project control and they serve a particular purpose at a point in time without due consideration of what the role really entails.

So, I can hear you think, 'what a mess', and you would not be wrong to think this.

This is because of several influencing factors that affected our project management environment in the mid to late 80s, as well as some strong institutional and contractual influences at that time (and even now).

In the early days, with not much software around – unless you were one of those people that worked on the Artemis software products (pre-PC ages, 'Palaeolithic' ages) – project control was performed by the planner in what was known as

PERT diagrams. Then as contracts and clients/contractors were battling it out in the courts for excessive claims, etc. (nothing much has changed on this even nowadays) some professional institutions got involved, but none on the planning side. Those days there were different prevailing conditions and various initiatives (e.g. Latham's (1994) 'Constructing the Team' and Egan's (1998) 'Rethinking Construction' reports) investigated the construction industry status and proposed a number of actions. These are not to be covered in this book, but I will attempt to provide some insight into why we are where we are now.

In my opinion the planner of the project has to be responsible for carrying out all the required project control processes. That does not mean that s/he sits in her/his 'silo' and performs only project control processes. The planner, as well as all project team members, work as a team and support / consult / are informed by the other disciplines within the project. But, it is the planner that combines, in the project planning software, time and cost and generates the project control information.

The point 'Who presents What and to Whom' will be discussed below but the planner has the responsibility for the project control information. Therefore, the planner is the project control person and performs the required activities on the project. The importance here is the integration of information in the planning software. This means ultimately no use of 'grey' IT by the other disciplines.

This is when considering the various organisational theories I can say that, based on the contingency theory of organising, 'there isn't a best way to organise'. So, there isn't a straightforward answer to whether or not project control is a department or something else. However, I will offer you some scenarios with their pros and cons so that you can make the best decision depending on your circumstances, as in the case of situational leadership.

1. Scenario 1 – A department of Planners/Project Controllers

In scenario 1 I will describe the case of a department of planners that do both cost and planning, which I will call project controllers, and generate project control information. The aim does not change and the department can be easily called Project Control department.

In this case all other activities are performed by other individuals, e.g. Risk Management, Document Control, etc., with the exception of reporting that needs to integrate information from the other disciplines.

Project controllers have the ability to work in the two different disciplines, can ask the correct questions not only in terms of the plan but also interrogate cost information, can critique the cost loading of activities, can extract information from 'Applications For Payments' (actual costs) and load these as actual on the activities.

The project controller, with the team, works on forecast and in particular the Estimate To Complete (ETC) values at activity level. It is a powerful combination and enables the direct integration of time and cost. It means we have a role that works closely, in UK, with the discipline we call Quantity Surveyor (QS) and at the same time enters the information in the planning software tool.

The integration, which is described in all text books, allows the team to look ahead in a dynamic environment and not just in a static / spreadsheet / contractual only based set of data. Whilst this may sound unrealistic if achieved this integration can bring an extensive advantage.

In terms of having sufficient resources to carry out these tasks, the right individuals should be engaged and where they are not or do not exist we need to develop them. There is no reason why we should be duplicating effort.

The drawbacks of this approach, apart from overreliance on the availability of these individuals, are around the interfaces between project controller and QS and the team's personalities (including those of the Senior Management). What I am suggesting here is for the QS / Commercial to allow another discipline to 'take over' their input and output on the project(s).

If you thought software and information integration has its difficulties, then you will have to add 'culture', 'behaviours', 'management attitudes', 'third parties' and the list of issues ends up being very long. Unfortunately, in the Information era we are currently in, it's the people factor and the complexity of interfaces that are the biggest challenge on the combined 'project controller' approach. To my knowledge, this is mainly an issue that I have come across in the UK. Two of the main reasons for this are the effect of institutionalisation and contract law.

2. Scenario 2 - A combined department - Planners and Quantity Surveyors

In scenario 2 I will describe the case of a department which includes the two disciplines, Planners and QSs who work together to combine information and produce integrated project control information. As previously the other activities, such as risk, document control, etc. are as previously performed by other individuals

.

This approach is the most common in the construction industry in UK. When the concept of team work and the interfaces between the two disciplines are functioning properly, this approach has many advantages. Advantages include (a) common line management, (b) common processes, (c) common tools, which leads to Planning and Commercial line management having the flexibility to allocate resources on the projects so they can support the Project Manager and the team. In terms of the project team, you have two heads – project controller and QS - instead of just one person challenging information received, reports issued, etc. The line manager has more flexibility to find the appropriate resources that will support the project team with no conflict between the two disciplines.

Another advantage, under common processes, is common reporting. Reducing the information integrity issue is a major advantage of this approach. Removing the use of different tools, with their respective outputs, is the advantage you get when the two disciplines are under one manager / in the same department. I have experienced the usual conflict between reports from the planning tool and the spreadsheets produced by the commercial discipline and trying to manage this is not a good position to be in. What you need to mitigate here is the risk that the interface between the two disciplines may cause problems for the Project Manager and the teams by issuing different reports or reports where the information is not integrated. The easier solution in this case is to get both disciplines to use the same tools, or at least tools that are interlinked. The latter, although it has been a target by a number of project management software companies in the last 20+ years, has to address a number of problems, which will be discussed in Chapter 7.

If we look at the drawbacks of this approach we will see that these are basically around people and attitudes, with some interference from the tools in particular 'grey IT' used for easy wins.

The biggest drawback, which was mentioned above, is getting the two disciplines to integrate without conflict. Most of the time the source of conflict stems from contractual process, which interface with cost control processes and which cannot be avoided but need to be managed. One of the two disciplines has to follow contractual terms and conditions to the letter and if the other one is not aware of these conditions then it gets frustrated and disengages.

This issue also becomes worse if the two disciplines report to two different line managers where a strict contractual approach is followed and there are demands for separate reporting. This forces individuals that carry out the tasks to follow different line management instructions, with all the unnecessary consequences.

### 3. Scenario 3 – Separate disciplines performing project control

In this scenario 3 we have project control viewed as processes / routines performed by a number of disciplines and with the output / reporting carried out by a separate reporting expert. We have an output here called 'project control reports'. What is considered here is a decentralised approach where project control does not sit within a specific department. Relevant processes are carried out by individuals, planners, commercial / quantity surveyors and others but the output / report is generated by someone else.

This may sound an expensive solution, especially when considering this approach in construction where the margins are very low.

*Are there advantages with this approach*?

One could say that there's a third 'eye' that receives and interrogates the information, asks the right questions, receives the appropriate responses and reports upstream.
I know, and I can hear you thinking, '*that's an expensive luxury that could and should be covered by management or a senior member from a discipline within the team*'. However, it is still an option that could be considered and responded to. For example, in the case of a client organisation it is possible that this approach can be considered viable, as the reporting function can carry out different reporting tasks as well. There's also some flexibility in the structuring allowing the two main disciplines to concentrate on their activities. However, this has to be weighed against the disadvantages.

*What are the main drawbacks with this option*?

- Ownership of information reported to the team and upstream:
  - Conflict between project control and PMs who are accountable for the information;
  - Conflict with other disciplines that do not see what is reported.
- Trying to understand what is reported and explaining output without prior knowledge of the projects;
- Undefined level of reporting requirements generates unnecessary level of workload;
- Dangers of 'grey-IT' used by the project controller, who extracts and manipulates information without the prior knowledge of the project team. Even the process of extracting information from proper project tools into grey-IT is prone to errors;
- The introduction of interfaces, as described earlier, causes complexity with all the problems;
- Then there is the issue of time available to fit all the required actions within the reporting timetable. This of course puts the pressure on the lower reporting levels and squeezing people's time to perform their duties.

### 4. Scenario 4 – Decentralised approach led by the Project Manager

In scenario 4 I will describe an alternative option but unless this is supported by all company levels it will fail from its onset.

I am talking about the option where the Project Manager (PM), in a centralised approach, performs the primary project control processes and is supported by others who provide the required information. This is different to the norm of a decentralised project control function where specialist individuals (as described above) are allocated to projects and could / are working with the project team in a matrix structure.

For example, the PM does his own cost loaded planning and receives from a central support function the actuals, from the site / design / procurement teams the progress information (if they are not entering this in the planning tool themselves) and then during the review date they discuss performance, resolve issues, review risks, future actions, etc. Some companies see this as the ideal. This approach actually does fit the scenario of managing a number of small projects and requires a good software and integration of information (through the standard project management structures and the software).

So these were the options with their advantages and disadvantages. However, it is unlikely that you have bought this book to be given more 'on the fence' answers.

Of course, the best solution is where time and cost and the respective project control processes are delivered by one person, the 'project controller'. However, and as mentioned above, this is not that easy and these rare individuals cannot be found easily, unless industry and academia do something about it.

From my experience, the most appropriate solution is that described in scenarios 1 or 2, where project control is performed by one individual or where, as a department, the two disciplines – planning and cost control (quantity surveyors) - report to one line manager. The line manager is knowledgeable of the requirements from both sides but does not act as a contract / commercial manager. Cost and planning activities are integrated. All work is supported by a common tool with no grey-IT (that is, no spreadsheets that record costs outside the planning tool).

The solution not only works at project level but it also works at client level. The reason for that is you have one interface with a number of other departments (in particular Finance). Information reported by the integrated project control is transferred to Finance and any questions are resolved very quickly. Accruals, 'GRing', invoicing strategy and other accounting requirements are not alien to the project team.

Having discussed the question whether '*Project Control is a department or something else*', in the next chapter we will explore the organisational structuring of the department. In some cases we will revisit the above, as the two are interlinked.

# 5. STRUCTURE OF THE DEPARTMENT

As discussed earlier there are two main functions that are considered as project control. These are planning and cost control.

The industry in UK has a view that is not necessarily the same in other countries, planning and cost control are considered separate functions performed by different people and belonging to separate departments. This is mainly driven by the institutionalisation of one of the functions. There are also other disciplines that are not considered part of the department, such as risk management, reporting and document control.

Based on my experience and the various positions I have held in the industry, I will describe the two views of structuring the department. I am talking about the wider view, as in one that is more integrating, and the narrower one, as in the more separatist one. We will look into the 'why' and the 'how' of each view and I will also discuss what the other considerations are.

1. The wider / integrated view

With this view, planning and cost control and all the relevant project management sub-processes and routines are:

a. Under the same department;
b. In many cases performed by one person.

A wider / integrated project control department, in addition to planning and cost control, will also include estimating, risk management, reporting and document control. This approach is especially supported by the improvements in the latest versions of project management software packages and the now possible seamless integration of functional information in the background.

With all these disciplines under the same department we have all project management support within one umbrella department with single line management responsibility. This allows for greater resource management flexibility, as in allocating the individuals from the department to projects, accountability, coordination, consistency and, above all, responsibility of the outputs produced. The latter could also include overall / company-wide reporting, as in rolling up information to higher levels.

In Chapter 8 we will see how the project control outputs are the cause for major issues and in Chapter 9 I will discuss the most appropriate approach to performing the project control tasks.

2. The narrower / separatist view

As is obvious from the heading, there are views, mainly in the UK industry, that support a complete separation between planning, cost and the other disciplines / functions.

In many cases these report to different line managers in a mixture of structures. Essentially you will have planning and cost control functioning separately and performing their own sub-processes.

The challenge here is to integrate:

a. The processes;
b. The outputs, and most importantly;
c. The behaviours of the people at <u>all levels</u>.

There are widespread examples of this approach. A startling example I came across was when I met a 'planner' who when asked '*how do you report on Earned Value (EV)*' in your organisation, he stated '*we / planners are not dealing with EV. The software picks up performance information, it transfers it to the cost software and reports are produced by the QS and the software*'. I do not think I can add more to this, but despair on the state of the project management knowledge and approach by those that implemented these processes.

Having said that, I will attempt to explain, but not support, why this view might have prevailed.

By keeping the functions separate you could be aiming to instil constructive conflict where individuals challenge each other on the information, outputs, performance, etc.
Well, this is a 'valid' view but ... to achieve this you need to have the following right from the beginning:

1. A well performing team environment with the appropriate behaviours at <u>all</u> levels;
2. Line management that respects and listens to each other properly;
3. Proper understanding of the project management processes and accepting the relevant standards;
4. Adherence to company requirements, which obviously respect and accept the project management standards.

In the separatist view of things, the other functions I mentioned earlier are not included in project control with estimating, risk management, reporting and document control under different departments.

We will see later on how integration of information in such an environment is a big challenge and conflict is unavoidable.

3. Other considerations

Obviously these two views on structuring the project control department are at a corporate level and in implementing them one must not forget that project control serves above all the project and the project manager.

From the above, it can be seen that the integrated view is much more beneficial to projects, though this is dependent on the project's size and the allocated budget. However, this can only be implemented on the planning and cost control activities, with the other functions either occasionally brought in or performed by various project team members. For example, the document control / support person takes the responsibility of compiling information for the project reports. In other cases, like for example risk management, the risk manager visits the project for the risk workshops, or the project management carries out this function.

In Chapter 9 I will describe further the benefits of the integrated approach to project control, but first we will look at the supporting documentation and how project control standard documentation contributes to the delivery of projects.

The PMP has to be a lean document and to refer to other supporting documents / standards, which are either developed for the project or are company-wide ones (see Figures 6.2 and 6.3, with examples of real project PMP). The example demonstrates that there should be a natural, almost hierarchical, link between the PMP and the other supporting documents / standards. In a similar way there should be a link between the Programme Execution Plan and the individual PMPs which describe the 'Why', etc.

The overall content should remain the same. References for more detail on the specifics should be made to the other documents. Ideally, references to other documents would be hyperlinked to increase the ease of access to them for readers, as well as it being easier to return back to the main documentation.

Figure 6.2: *Example of Project Management Plan* (PMP) *Contents page*

1. **Introduction**
   1.1. Purpose of Document
   1.2. Document Roadmap
   1.3. Business Need
   1.4. Project Objectives
   1.5. Brief Description and Background
   1.6. Location Map
   1.7. Project Performance measures and Targets
2. **Structure of Organisation**
   2.1. Introduction
   2.2. The Project Team
   2.3. Project Organisation
   2.4. Key roles and Responsibilities
   2.5. Management Process and RACI
3. **Team Directory**
4. **Programme**
   4.1. Project Implementation programme
5. **Commercial Management**
   5.1. Single Commercial Team Structure
   5.2. Financial system
   5.3. Procurement
   5.4. Commercial Tension
   5.5. Cost Management and Controls
   5.6. Target Reduction
   5.7. Commercial Strategy Document
6. **Project Processes**
   6.1. The Project Control Handbook
   6.2. Project Reporting
   6.3. Production Control System
   6.4. Programme Management
   6.5. Risk and Opportunities
   6.6. Design Integration
   6.7. Information Management
   6.8. Drawing and Document Control
   6.9. Drawing Review Procedures
   6.10. Request For Information (RFI)
   6.11. Change Control
   6.12. Method Statements
7. **Safety & Health & Environment**
   7.1. Objectives
   7.2. Emergency/Incident Action Procedure
   7.3. Construction H&S Plan
   7.4. Shift Activity Briefings
   7.5. High Risk Planning
8. **Quality**
   8.1. Strategic Quality Plan
   8.2. Photographic Defect Process
9. **Asset Maintenance Integration**
   9.1. Process map
   9.2. Process Guidelines
10. **Handover Strategy**
    10.1. Strategy and Processes
    10.2. Documentation and Procedures
11. **Appendices**

Figure 6.3: *Example of Project Management Plan* (PMP) *documentation structure*

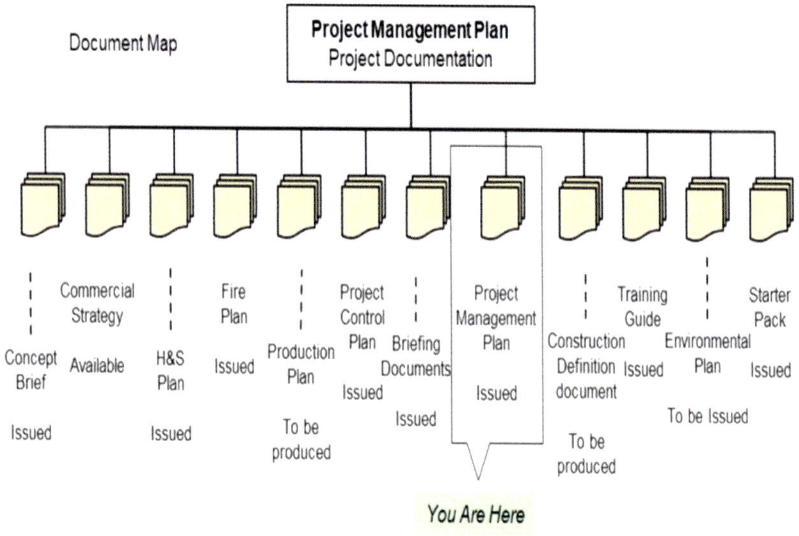


2. Template of the PMP

Very rarely in textbooks do you see example templates and that is for a number of valid reasons. However, I have developed and used on a number of occasions a template that allows project practitioners to consider all sections that are required in a PMP. What is described in Figure 6.4 is not restrictive but gives you a generic example of the PMP content.

In the template PMP I have followed, as one would expect in terms of 'good practice', a similar approach to setting up a WBS. It allows users to add, remove, replace at lower levels content according to the specifics of the project or the company requirements.

The various sections shown in Figure 6.4 respond to the 'Why', 'What', etc. that were described earlier.

The reader must not forget that with this template I am aiming to give some guidance, so it should be adapted to reflect the relevant project requirements.

Why do we, the project control, need to support the development of the PMP at its various levels?

Figure 6.4: *Project Management Plan* (PMP) *Template.*

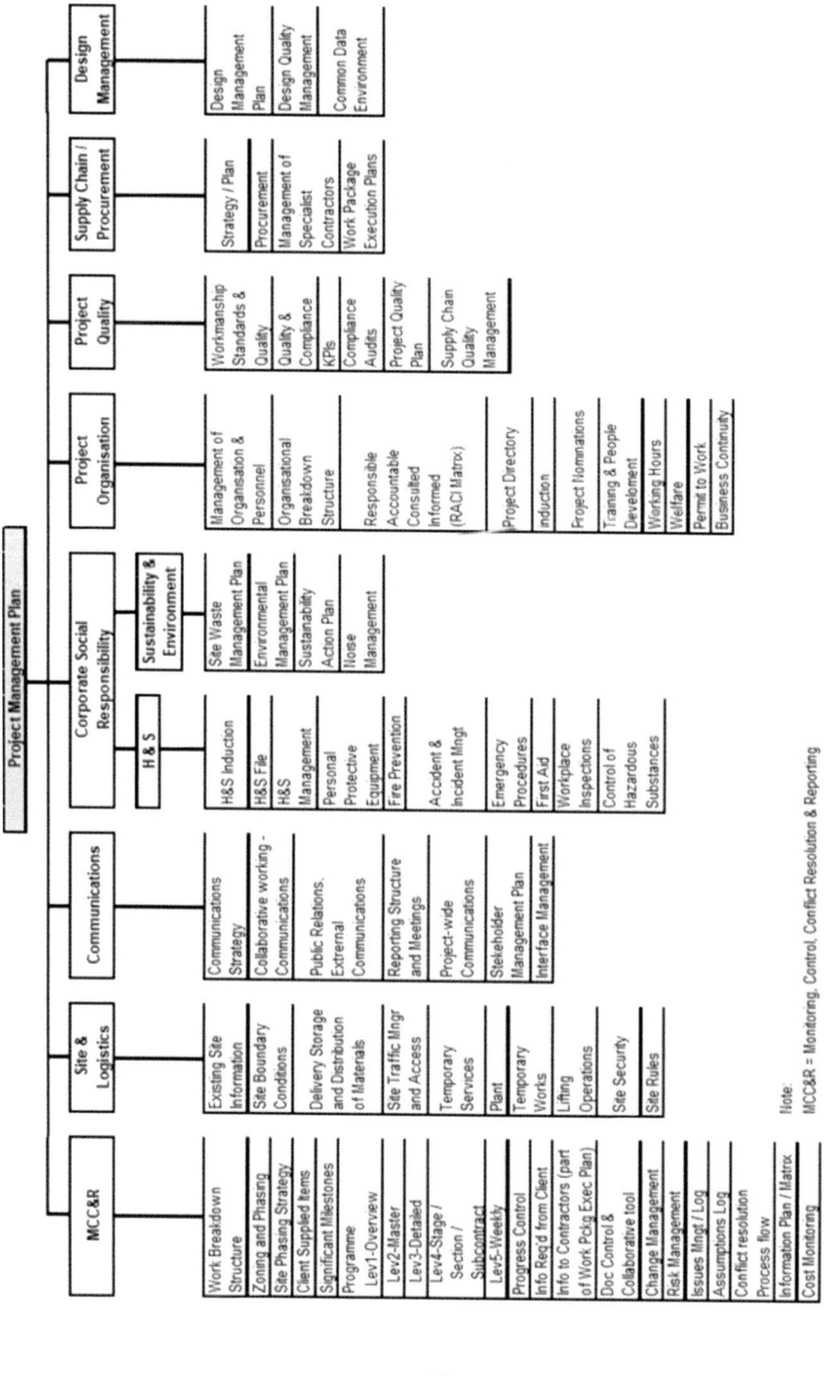

We have spent some time describing the PEP / PMP and the reason is that project control has to contribute to a number of its sections and therefore needs to be more than just aware of what is included. Accountability is with the PM but responsibility for parts / sections is with the various disciplines

3. Other project standard documentation

There are and should be other standard documentation that supports the teams to deliver the project. Unlike the PMP, I will not provide more details but just mention these, with a small exception made to include a particular document the Work Package Execution Plan (WPEP). The WPEP is not usually a standard but I found useful on a number of occasions.

Some of the standard documentation that needs to be available is:

- *The Project Quality Plan*: The project document that describes how the quality of the project outputs is checked and ensures that these meet the stakeholder requirements.
- *The Stakeholder Management Plan*: The stakeholder management plan sets out the preferred procedures, tools and techniques to be used in managing stakeholders.
- *The Risk Management Plan*: The project document that describes how the risk management process is performed, the roles and responsibilities, the criteria for successful risk management, etc.
- *The WPEP*: Earlier I mentioned how standard documentation follows a hierarchical structure with the PMP at the top of the pyramid. Following on from that, a standard document that should definitely be 'hanging' below the PMP is the Work Package Execution Plan (WPEP), see Figure 6.5.

  The WPEP should be issued for each package of work that is part of the project. This should be easy to develop right from the beginning of the project, and using the WBS helps the PM and the team think 'what' and 'how' needs to be delivered. The WPEP connects the higher level with the lower level packages of work. It covers and indicates to lower tier teams the 'Why', 'What', 'When', 'How' and 'Who' in a similar approach to the PMP.

  What this does is – partly – allow for the management of interfaces in conjunction with the delivery teams. They, in turn, can use this document to develop their PMP. Above all, and as mentioned earlier, it manages the effect of complexity from this interface and to be more precise, the effect of complexity through the contract management process.

  Again, project control plays an important role in the compilation of this document because, as you can see from Figure 6.5, it contributes to a number of the WPEP parts.

Figure 6.5: *The Work Package Execution Plan* (WPEP) - content page

## Contents



We have seen the other project standard documentation and now it is time to discuss the one that is set up and used by project control.

Yes, I am going to talk about the Project Control Plan (PCP).

2. The Project Control Plan

In this part I will describe extensively, what the Project Control Plan (PCP) is, its objectives, the structure and a number of other important topics. I will also include a number of examples.

1. The Objective of the PCP

The objective of the PCP is '*To provide adequate guidance to all working within the project / organisation through the project controls framework regarding the setting of cost loaded programmes, monitoring progress, outputs and all relevant project management 'control' processes*'.

2. What does it address?

The overall approach should be based on the principle of integrating time and cost to an adequate level in order to:

- Know what must be done (i.e. the plan).
- Know what has been done (i.e. measuring performance).
- Know what has gone according to plan, what has not, why and what remains to be done (i.e. monitoring and comparing performance against the plan).
- Support proactive management actions and decision making.

3. What is it and what are the different levels?

There are two main levels of the PCP. The company, or it could be said Programme (as in programme of works) level and the project level.

The structure of the document does not vary between the two. The only difference is the level of detail included in the PCP for the project.

As is described above, the PCP is a guide and basically indicates how the team addresses the project control set up requirements. Therefore, it has to be brief, concise and easily accessible.

The Company level PCP indicates and sets higher level requirements. For example, it contains the company WBS and other structures; it links to company-wide reporting processes and requirements, etc.

The project level PCP describes the lowest level requirements and what the teams have to do. It contains the appropriate level of information and actions that will act as part of the project induction, quality assurance and in many cases as a guide to lower tier subcontractors, suppliers, as it really should be part of tender documentation, etc. It will also be the link to other documents such as the Work Package Execution Plan (WPEP) mentioned earlier.

Above all, the PCP includes all the project management standard structures that are to be used in the project or at a company-wide level. That means, it presents the WBS, OBS, CBS, coding and other required information that structures the project control data.

An example of both levels of PCP can be seen below in Figures 6.6a and 6.6b as well as in part 9 below.

4. How is it structured?

When considering the structure of the document we need to understand the importance of setting out a clear objective, given the use of the PCP as a key document. Following on from that, it needs to reflect and include the company specific requirements.

Then we have the detailed set up, starting with the description of the standard project management structures. It needs to describe the project, or company-wide, WBS, OBS, CBS and other coding that enables the integration of information.

Following on from that, we have to include the descriptions of the various project control processes. These are the planning and cost control, progress monitoring, risk management, etc. The details for each process, as I mentioned earlier, should be included in the Appendices and this should be of particular interest to the individual disciplines.

Figures 6.6a and 6.6b provide an example of a multi-million pounds programme level PCP and give a good indication of the document structure.

Figure 6.6a: *Example 1 of Project Control Plan* (PCP) – *Contents Page*

Figure 6.6b: *Example 2 of Project Control Plan – Contents Page*

**Table of Figures**



**Appendices**

5. Where is it distributed?

I hope that you will agree that this is a very easy answer, and that is everywhere and to everyone.

There should not be a restriction to the circulation of the PCP. All internal and external parties to the project have to be informed, understand and implement the instructions given by the PCP.

There isn't and shouldn't be an issue of Intellectual Property (IP), with the exception of military and other such projects, because the PCP sets out obvious project control guidelines that have been part of the project management theory for a number of decades.

Additionally, when the PCP is issued to the contractor(s), external designers and second / third year suppliers, it allows everyone contributing to the delivery of the project to be aware and follow the guidelines. Also, this contributes to a more collaborative environment. There cannot be anything more important than using a document that describes 'control' to support change in 'behaviours'. To give an example here, the PCP will have to describe how the change control process is to be carried out (not the contractual one) and how processes, systems, and people will have to follow this.

There is also the fact that by distributing this to all, it allows third parties to see how they will collect and issue information to be used by all relevant corporate systems. This, in partnering environments, allows for integrated processes and a true 'Bottom-Up' approach to integration of information.

6. How are these rolled-out?

As with any other standard documentation that is to be followed by a wide audience, the PCP has to be rolled out in a structured manner. Irrespective of the level – whether company-wide or at project level – the Project Control department (or the PMO) has to implement a communication strategy for the roll out of the PCP. The usual pitfall in which PC departments fall is that they consider the PCP as only a departmental document.

Workshops, lunch & learn sessions, knowledge transfer sessions or similar events are a means of sharing knowledge and it is preferable to do this in a multi-disciplined environment. Communication about the PCP cannot and must not be a one off event.

Firstly, the PCP has to be easily accessible. It needs to be available in the company, Project Control / PMO and / or on the project intranet page (if there is one).

Secondly, frequent reminders should be issued about specific parts of the PCP. I know that this sounds like a luxury in the busy project world, but it does not take long to extract a section of the PCP and circulate it in a Project Control communications flyer.

Thirdly, as changes and updates are happening these need to be widely publicised using the project control systems of communication, which we will explore in part 3 below.

Finally, the PCP is a project control document but it needs input and consultation from other departments and definitely from the PMs. It cannot be forgotten that it is part of the PM's armoury and one of the project success' criteria is project management success (Collins & Baccarini, 2004).

7. Forever draft?

Another point that needs to be addressed is if the PCP can ever be considered as complete and therefore no further amendments, additions or changes are needed. As with every document that needs to be kept up to date, similarly the central / company PCP has to remain in an approved but 'draft' state. That does not mean it shouldn't be followed, but it should be taken as something that has to be updated as and when required.

In the case of projects, not those mega ones that take several years, the PCP will mostly use the company-wide PCP as a base, expand and define in more detail those parts that have to be *'fixed'* for the project and then really not change much.

So, it is forever draft at company level, but pretty much fixed for normal duration projects.

8. *'Loose straight jacket'*. Why?

We all know that no project or Project Manager are the same. In some cases even the same PM follows a different approach from one project to the next. This is not wrong and for that reason project control has to allow for a degree of flexibility.

However, there are elements of project control which cannot allow for any changes / alterations without consultation. An example of that will be the WBS and the other standard structures. So, the PCP has to define the parameters within which the PM and the teams will have to function/perform and by doing that it becomes a *'loose straight jacket'*.

*'Loose straight jacket'* is an expression that I have used on a number of occasions to describe to project practitioners the environment they will be operating in with project control. It allows the PM and the teams the required

flexibility whilst at the same time remaining the guardian of the project management standards. The department has to carry and balance this tall responsibility.

So, here you have it: the Project Control Plan has to be seen as a '*Loose straight jacket*' and actually to be followed for this reason.

9. Samples of the PCP:

Figure 6.7 is a PCP from a specific project in the infrastructure industry. Figure 6.8, is an example of a company-wide PCP that was rolled-out in the utilities industry. It covered an annual delivery of approximately 500 projects. This will give you an indication of the type of '*Loose Straight Jacket*' that had to be applied to a population of approximately 90 PMs.

Figure 6.7: *Example 3 of a Project Control Plan* (PCP)

[redacted] Project Control Handbook - Local Implementation Plan

**Contents**



**Table of Figures**



**Appendices**

Figure 6.8: *Example of a company-wide Project Control Plan* (PCP)

**Contents**



It has to be noted that in both examples above the content is not restrictive, and these reflect the requirements at that particular point in time. For example, in Figure 6.7 the PCP did not include the OBS because this was part of the PMP developed by the PM.

In the earlier chapters I described the basic supporting documentation. However, as I have previously demonstrated, we cannot ignore communication.

One of the areas that project control tends to forget is that of communicating, not only producing reports, but generally sharing what the group is doing, why, how and when. The latter is not covered in reports, which is evident as part of the PCP.

In part 3 below I will suggest some of the means that can be used by project control to communicate with the other disciplines, the project teams and in many cases (as in client organisations) with other Directorates.

## 3. The Communication tools

The power of information is useless without good communication and dissemination.

Project control generates reports for projects, programmes of work at directorate level as a direct output from the work on projects and the software tools available. However, in addition to those, there are a number of other activities that are carried out by the discipline as described earlier. Some of the means that can be used to communicate these to the wider population are:

1. **Flyers**. Document controlled announcements describing / addressing a single issue and providing clarity.
2. **Intranet**. Announcements on an intranet page either dedicated to Project Control (or PMO) or the Project.
3. **Viz-boards**. Electronic reporting boards, otherwise known as viz-board. (Please let's move on from paper to soft reports. It is about time we bring people along to the 21st century). I think, as far as project control is concerned, this is self-explanatory. Project control starts with information entered and extracted from software tools. Therefore, it is very easy to use viz-boards and produce soft copies of reports that can be presented on screens in the project or company offices.
4. **Lunch & Learn** sessions. If the issue, process, change to be rolled out is much bigger then this requires more time and face-to-face sessions. At these sessions the group can be taken through the subject in detail and there should be extensive Q&A.

Coming back to 'what counts as  a bigger process' to be rolled out, I am talking about using such sessions to describe subjects like 'Common Data Environment', company-wide Risk Management, Data integrity, etc. There are some subjects that require a concerted effort, not only from project control but from the wider project team and in many cases the company.

5. **Knowledge Management hub**.

This topic, although not new in the industry, has only lately found its way to the top of company and other organisations' agendas.

It is a massive area, so I will only touch on how the project control department, or how the PMO / project control, in a wider role, can incorporate this into parts of the deliverable services.

Within the banner of 'PMO Memory', which includes areas such as 'Guidelines', 'Lessons Learnt', 'Viz-boards' and 'Analytics' we can also incorporate Knowledge Management. Figure 6.10 gives a mind map view of these areas that can become part of the PMO Memory.

The content of the first two areas, as is obvious from the title, contain instructions and capture lessons learnt. Analytics and Viz-boards are all about reporting and capturing and communicating the information from projects.

On the Knowledge Management (KM) part the Project Control department, within its capacity, can easily become the KM hub. This can be done by simply incorporating a number of areas, as shown below in Figure 6.11. Through the KM hub project control can integrate areas and deliver another service to the Project Managers. These do not have to be only process oriented, as can be seen below.

6. **Lessons Learnt** (LLs)

In many occasions, Lessons Learnt (LLs) become part of project control. The basic concept of LLs is that these need to be a continuous process and should not only be carried out at the end of the project. PMs need to lead the LLs sessions and capture events and project information that will become the basis of learning on a continuous basis. The process is described extensively in the major project management documentation so I will not expand here any further. I will only provide you a practical example of capturing LLs continuously and the approach that can be used by adding the power of the available software tool (see Figure 6.9 below)

Figure 6.9: *An example of a Lesson Learnt approach using a specific software tool*

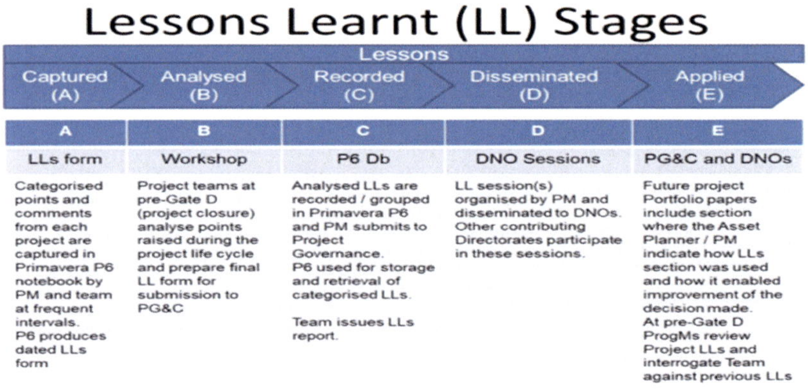

Figure 6.10: *A mind map indicating areas that can be covered under PMO Memory*

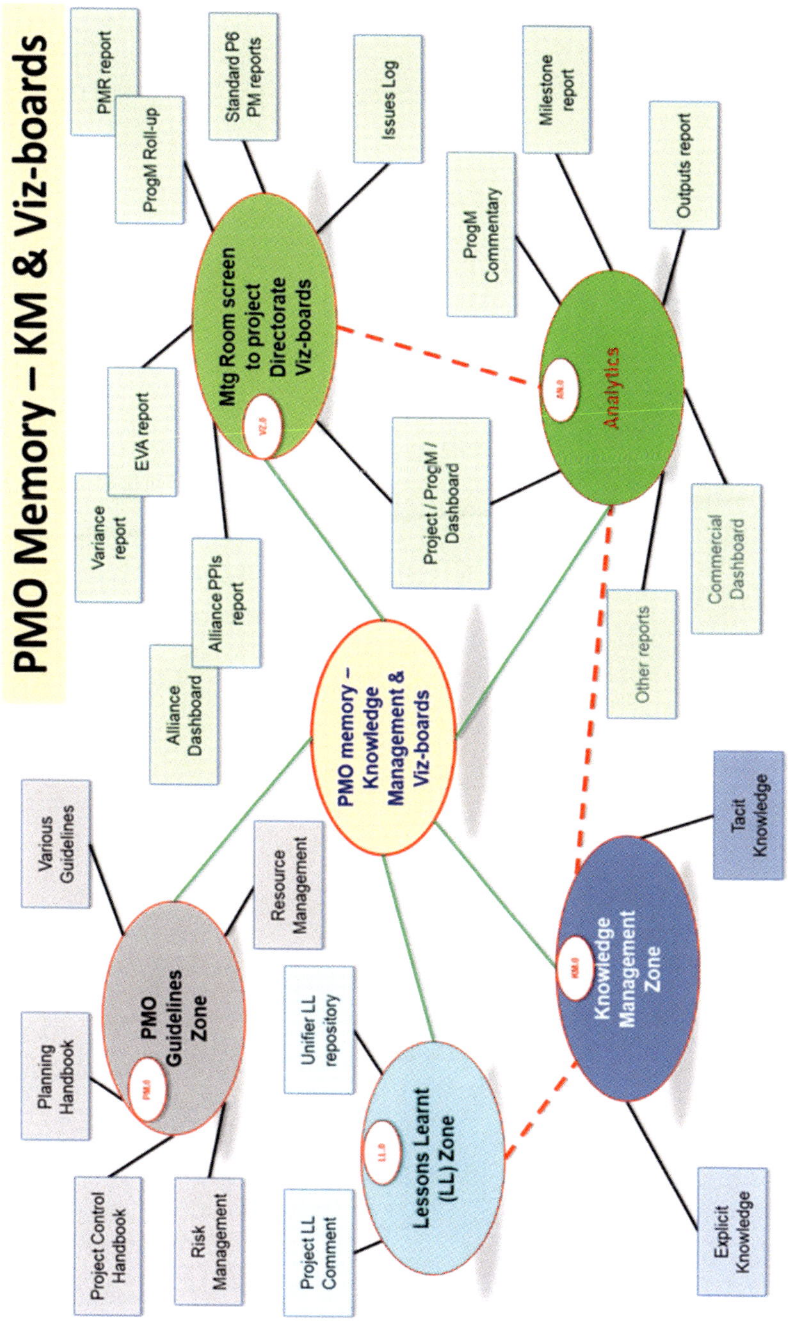

Figure 6.11: *A mind map indicating areas that can be covered under Knowledge Management*

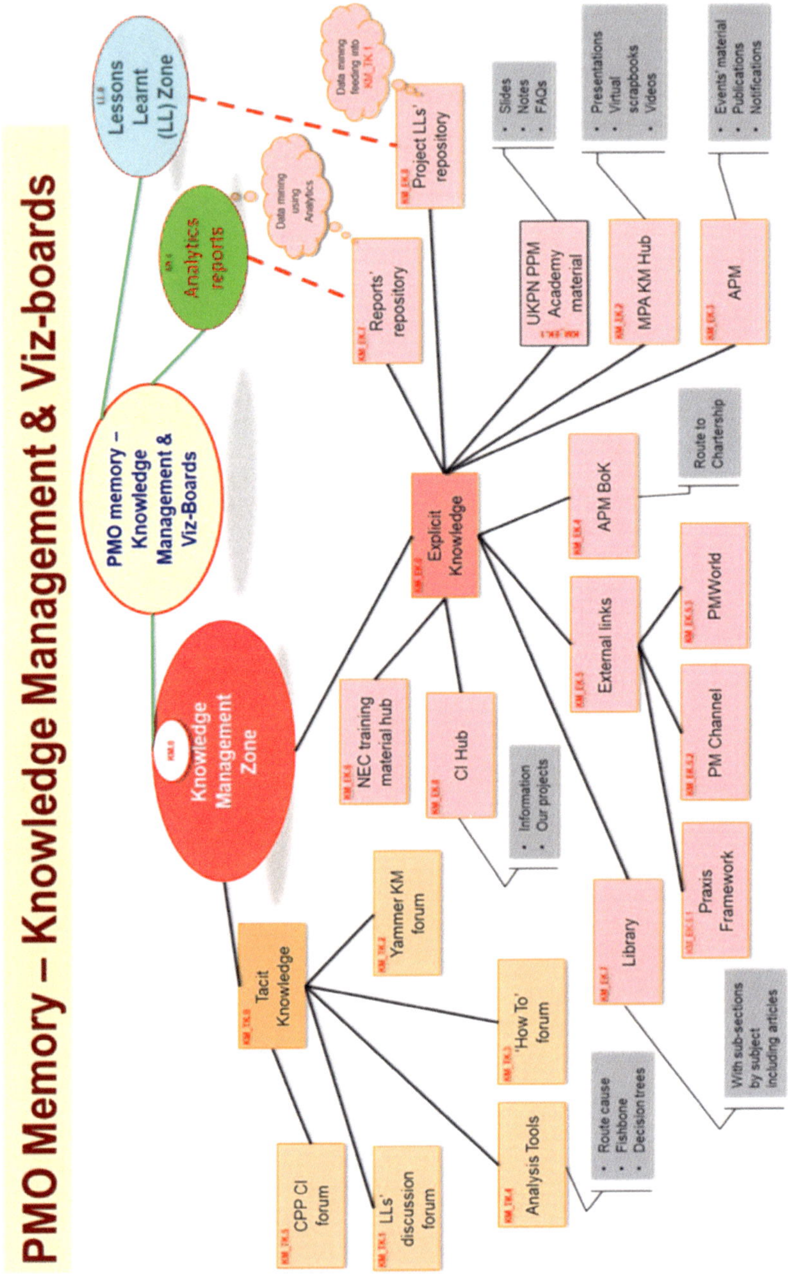

# 7. PROJECT CONTROL SOFTWARE TOOLS

As described earlier, in the project control objective chapter, project control is all about integration and this can only happen by integrating software tools.

In the early days of project control there was one prevailing software supplier who was providing the basic database and that carried out the 'precedence' programme calculations when you entered activities and logic links. That is... the software produced the Critical Path Analysis, giving you the programme and then everything else was up to you, the user, to design and set up. The database was a Relational one and with some basic knowledge and good thinking, you could set the:

- Cost database;
- Reporting database;
- Drawing register, as a database;
- Graphics database;
- A number of other interesting set ups.

All you had to think about was your 'key' field(s) and how to interlink the databases.
Then with the development of PCs, and some disasters in between, by trying to use the 'C' program language, things started moving faster with more users and more accessibility.

Then we had the advent of Enterprise Resource Planning (ERP) software 'beasts' and at the same time the planning software companies started moving towards integration / merger with the ERP software tools.

Now we are in a period of further consolidation, where smaller specialist tools are built to easily integrate with major planning software packages and / or ERP software.

However, whichever way you look at it, the project controller, has to be able to grasp how systems integrate and what the basis is for this. And knowing how systems integrate is how the battle of the project programme/schedule been the driver and not other software tools is won. Using the saying 'if you don't know your history, you don't where you are going' and thinking back to the early days of the scheduling software years and the profession, the basis of the software was the logically linked and cost loaded schedule. I am referring to the times when two of the three 'gold' triangle variables 'time' and 'cost' were not separated and project control was handling both.

The figure below (Figure 7.1) presents the 'Integrated Project Control System' and the basics of what needs to be considered on implementation.

Figure 7.1: *Diagram of a proposed Integrated Project Control System and some of the areas that need to be considered when setting up a Project Control system*

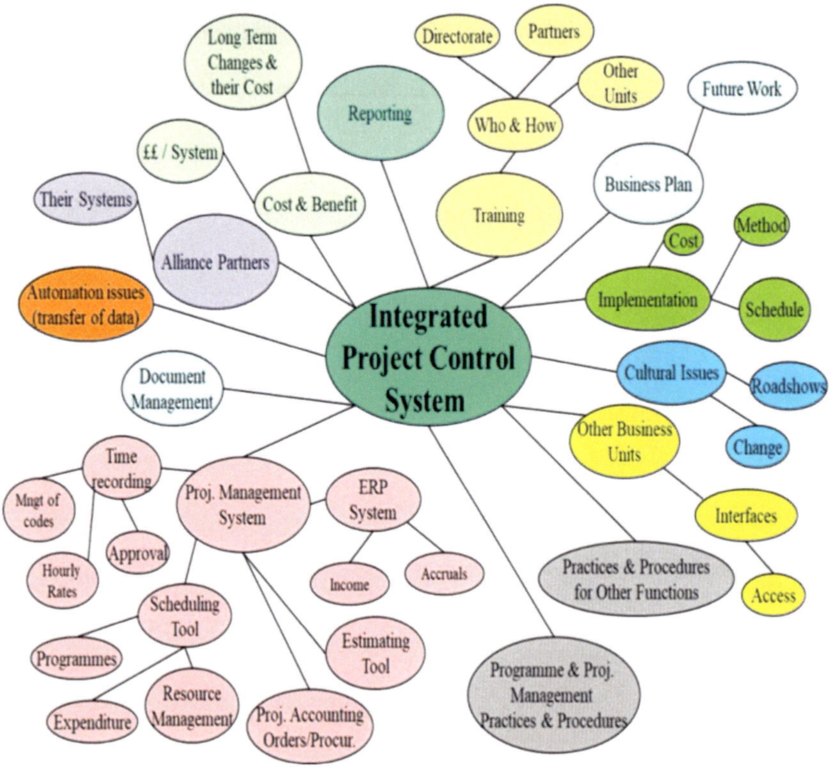


I will expand on some of the less obvious nodes:

- *Project Management System*
  This describes what needs to be considered and delivered from the scheduling tool.
- *Programme & Project Management practices and procedures*
  Any system has to consider the company programme & project management practices and procedures and accommodate the requirements.
- *Training*
  There will always be training requirements for those in the periphery of the system and at various levels.
- *Cultural issues*
  This is something that definitely has to be considered, as both culture and behaviour will influence implementation as well as continuous operation. There will always be those direct or indirect users who will try to avoid

following the required processes. This means that as project control you will need to 'police' the system.

- *Cost and benefits*
  Any project control system will require a cost and benefits analysis. You will have to identify tangible and intangible benefits. The challenge will be to make the intangible benefits understood so that you can support your business case. Just a note on this, there will have to be continuous improvement thinking as not all requirements could be implemented at the first stage. It will also require continuous business support as this will be a long-term project with staged deliverables.

Having discussed a bit of history and the integration of the systems that will deliver project control, we now need to look at the integration of the systems in the modern era and examine issues such (1) 'integrated or separate software tools', (2) 'what to avoid', (3) 'one or more software tools'.

1. Integrated or Separate software tools - 'Systems thinking'

My starting point is that project control systems have to be combined, given that information residing on separate tools causes confusion and is the main reason for generating complexity.

The usual separation is between time and cost. Although tools are capable of holding both, people may prefer to keep costs in spreadsheets or reinvent the wheel, as this is easier than understanding how things work.

Another example of keeping data separate is when monitoring of project risks is done separately from the project programme, using spreadsheets again. Even though combining the two in the same software tool is a recent development, in earlier days it was still possible to keep track of risks combined with the programme, but the willingness was not there.

Keeping information separate causes not only data integrity issues but introduces complexity, causes disruptive conflict between individuals and teams, confusion from reports that are generated and issued using separate sources and uncomfortable situations throughout the project and the other levels. And this is only if you consider one project. The problem becomes even greater if in an organisation – client or contractor – every PM and every discipline (head or individual) have their own 'system'. Rolling up and interpretation of information is near impossible. The problem becomes worse again when the colleagues in finance have to impose monthly cut-offs that are not aligned with those of the projects (or really the other way round). Here also I will not expand on 'accruals', 'value of work done', 'actual' (as in those entered in the planning software tool).

Project information has to be, as much as is possible, integrated in one or more software tools that are interlinked and the basic link would be the 'Project unique ID'. This integration is not only relevant for the project data but also for linking up-stream to finance, sideways to other departments, e.g. Procurement, as well as downstream to contractors through the WPEP (Work Package Execution Plan) discussed earlier.

The integration of the tools in some cases is much simpler than the integration / understanding and agreeing of the terminology to be used. This becomes a greater problem when contractual terminology, with its lack of flexibility, is imposed on project teams and tools.

All the above I believe give you rationale for integrating all project information and as I explained this has to be done through interlinked software tools. Now, the leaner the interlinked systems are, the less redundant duplication happens and therefore this is better overall. For example, planned expenditure does not have to be reported in both the planning tool and the cost management tool. Which one leads is a company decision. My preference would be that all planned cost information comes from the planning software tool. Estimates in a cost management tool (not linked to the schedule) provide worse estimates than using logically linked and cost loaded activities and a properly reviewed programme. Also, the cost loaded activities mostly have to deal with and are connected to resources, therefore outputs can also be used by others to manage departmental requirements.

Figure 7.2 describes an integrated system that I drew at a Utilities company and was implemented as part of a transformation program. It gives an indication of what can be achieved when one considers the whole project control approach under the systems thinking principle. It provides a view of how the software tools can be integrated and at the same time support a company-wide ERP system.

I will expand on some of the contents of Figure 7.2 considering the basic project control software tools that were interlinked and also describe the outputs, shown in Figure 7.2 as areas A, B, etc.

The whole software system solution was based on the Oracle suit of tools. Just to make it clear I did not /do not have any commercial benefits linking back to the software company. To put it simply, I used software tools from other providers and set up exactly similar systems. One tool that was always in the middle of the systems diagram was the planning software.

The central software tool was the Primavera P6 planning tool and all projects entered on the tool were logically linked and cost loaded. The Project Control Plan described how this was happening and set out the standards, like taking baselines, progress, change control, etc. to be followed. The basic structures,

WBS, CBS, etc. were also set up and allowed linking to the Cost / Contract Management software – Unifier – and the Estimating software – CostOS. The basic link between the three tools was the project ID.

Unifier, the cost / contract management tool, monitored actuals and change as described by the company project contracts awarded (which in this case was NEC). Actual costs, again in this case because of the ERP system requirements, were uploaded to SAP and then transferred to P6. The two tools can link directly and exchange information without any other interventions.

The interface between the software tools happens through the standard project structures. Therefore, the importance of structures, as has been highlighted throughout this book, is paramount and becomes even more important when these tools have to interface with ERP software tools.

Now within the planning software we had the ability to generate the risk registers (see Area B). Following the standard process, risks and opportunities were identified, allocated probabilistic data, mitigating action and all the other relevant information. Most importantly, these were linked to activities. This enabled not only ease of monthly monitoring but also generated quality reports which monitored compliance. This is something that no spreadsheet-based risk register can yet perform. On the reporting side (see Area A), aside from the usual risk registers, it was possible to generate a schedule based forecast of risks and opportunities and roll this up to Programme, Portfolio and Directorate level.

The combination of planning and risk in one software tool gives extra capabilities to project control to provide an even better / holistic view at the various reporting levels. In addition to the reporting, it also allows the PM and the team to work even closer without conflicts on data integrity and accuracy, as long as these interfaces are there from the beginning and that people talk to each other.

There's so much that can be said on having integrated systems and the real-life example I gave above, but saying more might sound like a software sales pitch and this is definitely not the intention. The fact that I wish to highlight is that there is a lot more that project control can achieve with integrated software solutions. However, this does require good early thinking and following the proper project management structures. Also, it should be said that other software proprietors do have planning tools that integrate with cost tools, e.g. Welcome with Open Plan and Cobra, Microsoft, etc.

Figure 7.2: *Example of an Integrated Programme & Project Management Control software system.*

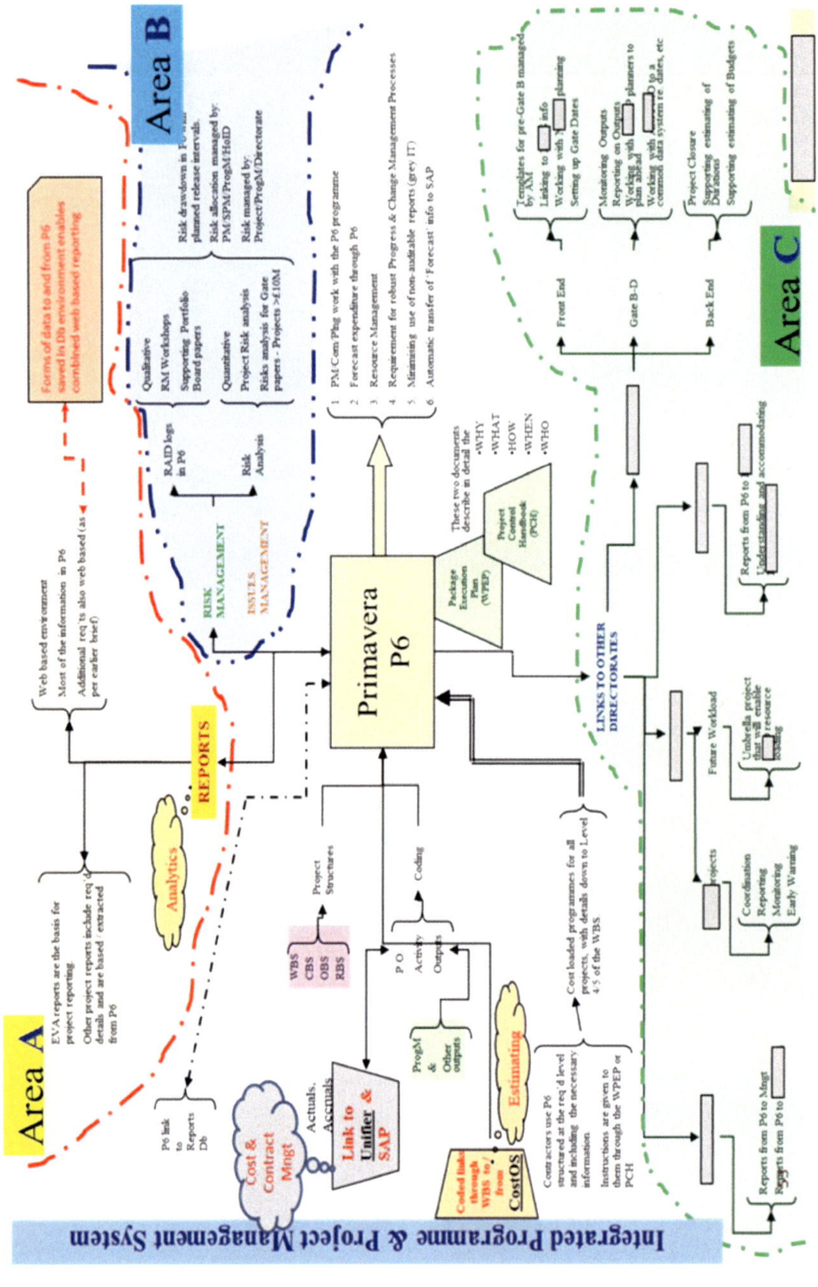

2. What to avoid

There are a number of points that you need to be aware of and avoid when 'designing' or even working with project control software packages.

You need to avoid 'department-centric' systems. By that I mean you will need to consider the parties that will work with the tools. This includes the functions or departments you are supporting. You will also need to consider whether your system is for one project or more than one, one team or a number of them and if there is a requirement for transferring information to other software packages for company-wide reporting.
Then you need to avoid manual interventions. Your aim will be to minimise these between different data entries from the various disciplines. If templates can be used then it is advisable to set these up from the beginning and issue to those that will be providing information at the requested intervals.

Another challenge you will face is that of exporting information to other software tools. Usually this comes with the rationale 'I need to produce a better report', or 'I need to capture *some* of the information you have'. Bottom line, this is a recipe for disaster and data manipulation. Anything that is exported cannot be considered as 'robust' and it is worth trying to resist this. Therefore, you will need to be setting up for 'reporting' having the 'end' product in mind and that is: to provide decision making information without the need for further manipulation.

Then you need to think of the external parties to the project / the system. Interfacing with external designers, contractors, suppliers is extremely important and therefore you cannot afford not to consider their needs. There should be a number of documents – from the contracts to Work Package Execution Packs – that you have to influence and to see what requirements to include from the project control systems side. It is best to avoid a situation where, every time you receive information from external parties, you waste time deciphering the information. So, in addition to the project teams you have to consider the structure of the system(s) in terms of the externals. Your best ally on this will be those external parties that will work in the project.

Above all do not compromise in the use of the standard structures. These will interlink your information, either with a hard or a soft link and will be used in the future for reporting purposes.

3. Using one or more software tools

This choice is not really just yours since the Information Systems department, operational expenditure per annum, support and other issues will affect your Project Control requirements. However, as project control, you need to have direct input on the department requirements, the tangible and intangible benefits, the upwards and downwards requirements for information, as well as other information, i.e. procurement, treasury, etc.

The basic approach these days is for the planning software to interface with the cost management software and upstream to the ERP software. Therefore, it should be taken as given that there will be at least two software packages. The interface between the tools, the processes and functionality needs to be addressed.

Below I set all the criteria you should consider.

a. How

This has been discussed extensively earlier and the answer is through the use of standard project management structures as the basis.

What you need to be aware of and consider is to allow flexibility, especially in the case of a client environment, for multiple types of projects. This means you need to consider a level of flexibility for the WBS and CBS. The reason is that the size of project varies and your project control 'straight jacket' cannot be very tight. Layering the structures to capture majority of projects is your challenge.

Here is a hint of your first challenge on the 'How': If your project control system is to link up to an ERP system you will not be able to have a number of very small projects under an 'umbrella' project, unless there is extensive software interfacing. What do I mean by the 'umbrella' project? Generally, for small projects that last from a couple of weeks to a few months (with small expenditure) it is practical to use one overarching project in the planning tool to capture the works in these small projects as activities. Work-arounds are possible for the WBS and the other structures, however, the issue linking upstream to the ERP will have to be considered as these systems cannot capture information at activity level.

Something else to be addressed will be: who does what and within which tool.

Additionally, you will need to ensure that planning interfaces with cost control, if these links are not being managed by the project controller and that information is entered only once. Then there's the soft interface between the tools. You will have to make sure that during progress reviews everyone looks at the same information even if the data resides in different software packages.

Regarding other information, both the modern planning and cost software packages allow, through the use of 'User Defined Fields', to add codes against activities, projects, resources, etc. and to group data for reporting purposes.

b. Spreadsheets Vs Databases

This is a very big IT subject and I do not intend to extensively describe the advantages and disadvantages of each. I will restrict the content to only the issues that affect project control; the benefits and for which discipline; the drawbacks and for whom or specifically which sub-process (or for a more academic terminology, which sub-routine).

On the cost control side, in the early days of project control, as well as in the early 2000s, there was not an easy to use cost control tool. Project controllers that dealt with cost and commercial information had to use spreadsheets. These provided, and still do provide, the user with the flexibility to structure, record and report according to the stakeholders' requirements. However, it didn't provide database characteristics. So immediately you have a group of people that have used and continue to use spreadsheets because '*that's the way we've always done it*', or '*we need the flexibility*', or '*we can do better graphics*', among other reasons. At this point in time, with software that can cover all this, I would suggest that there is no need to continue using spreadsheets for cost or contract control.

I know what you are thinking, and yes there are *'professionals'* out there that use spreadsheets to do planning, calculate backward and forward pass, calculate Earned Value and its Indices. There isn't much that one can say apart from what a wasted effort.

Just imagine the consequences and the lack of quality from project control services that are provided with planning and cost control conducted using spreadsheets. It is so 'flexible' that it lasts only one reporting period, then next month the process is repeated. And just to be absolutely clear I have witnessed this on 'major' programmes of work.

There are things that you can do, in order to help you, by using spreadsheets, but in the 21st Century you should be looking to use established software tools to do project control and these are only those based on a database.

4. Other reasons for integration

There is also another driver for integration of planning and cost via the proper software tools.

You are likely to have heard by now, if you are keeping up with developments, about BIM (Building Information Modelling) or 5D delivery. Just to make something clear, this is not a new concept, I came across the 3D 'planning' concept back in 1996. Back then a major International consultancy had put together a software package which integrated cost loaded programmes to CAD drawings and presented how design, as well as other processes, were informing the programme and vice versa. This has resurfaced, under the acronym BIM, because it is obvious that there are massive efficiencies when integrating programme, estimation, design (especially in 3D) and then monitoring these.

There is a lot to be said regarding BIM, but this is not within the scope of this book.
Also, the UK Government has made the implementation of BIM compulsory for a certain size of government projects. Therefore, companies will have to get on board with using proper tools to do BIM and project control.

The question is, can an organisation really comply with BIM using spreadsheets and project control software that is not integrated?

5. Integration with other tools

In this chapter we investigated the integration of the project control software tools and why specialist tools should be used rather than spreadsheets. Another reason is the integration and interfaces with the company Enterprise Resource Planning (ERP) software. This is obviously something that has to be considered at company-wide level. Figure 7.2 is such an example where on the left-hand side you can see the interface with SAP. Again, SAP is not to be taken as the only ERP software around, but at some point you as the project control department will need to consider the optimal way to transfer information at the required interval to other departments and/or levels.

From the project control side nothing changes in terms of the 'how'. The standard structures should be used, and in particular the WBS. ERP software tools have the WBS as a standard. All you have to do is convince the Finance department that they should agree to that. I have to admit that this is not an easy task and you have to be well prepared for this. It will help if you have WBS and CBS agreed and integrated and then present your case. Our colleagues in finance tend to like structured information that can be:

1. Issued in time at the month end.
2. Clear – Labour, Plant, Material, Overhead.
3. Accurate about the level of accruals.
4. Generally auditable and align to Balance Sheets.

ERP tools will give them this information. What you have to do from your side is have an understanding of their requirements and the importance of company accounts. If you think that with project control you are putting the project team in a loose straight jacket, then finance, as a department, have to operate in a very tight one. So, they do deserve some sympathy and consideration. However, something to be made clear from the beginning will be that project expenditure and financial accounting are different to a certain degree.

There are a lot of other benefits from the integration of project control tools and ERP tools, especially in a client environment. Consider the increased ease of procurement / supply of equipment to high level treasury activities when they are notified early for purchase of equipment that involves foreign exchange and hedging to avoid the risk of high currency costs on your project(s). This is easily done by giving them a list of your project(s) procurement milestones (with expected down payments). Think also on how project risks can be escalated to higher (executive) levels.

Together with the many benefits you need to be ready for the great pains as well. Defining, organising, setting up and rolling out ERP systems is not easy. That is why you hear and see that many Transformation programmes have been unsuccessful. It is not my intention to expand further on this but there is one point I want to highlight. As project control you need to have a holistic view and consider the whole system and its stakeholders – from your relevant delivery colleagues and their requirements for project information, to finance, procurement, governance and obviously the company hierarchy.

This is a very large topic and requires a book on its own. What I have tried to do here is to describe some of other major factors that require the integration of project control processes and software tools. I have also given you an indication of the importance of your role as project control. It's not rocket science, but simply helping people coming together, being understanding, considerate and willing to see the whole picture. The concepts are decades old but it is the interfaces that cause complexity and you need to understand how are these generated in order to be able to manage the effects of complexity on project or programmes of work.

# 8. PROJECT CONTROL OUTPUTS

This is the chapter where I will attempt to describe one of the most controversial topics in Project Management and Project Control. Yes, I am talking about reporting and the outputs from the control operations / processes.

*Why controversial?*

Well, experience has taught me that every single person involved in a project has their own views of the required report(s). Despite basic standard reporting being around for 60+ years now, reporting is an area where even the basics of Earned Value Indices are distorted, even after extensive presentations. I have even come across individuals trying to redefine Cost Performance!

In reporting, the project team and project control have to deal with intricate problems that involve:

a. People issues;
b. Organisational related issues;
c. Technology related issues;
d. Process related issues.

All the above deal with interconnections and interactions of structures, systems and people.

*How do the issues around outputs / reporting occur?*

a. We have reciprocal interdependencies. That means you will have to consider the interdependencies between the parties that come together to deliver the project (inter-organisational ones, as well as external to the organisation);
b. Information flow is bi-directional, based on the formula ($n^2$-n) {where n=number of info nodes / parties}. As an example, for this formula, theory says that if there are three parties to the project (n=3) then there are six information flows and if there are four parties then there are 12 information flows;
   This shows you the intricacy of communication / information flow in projects.
c. Filtering / noise in the reports, introduced for various reasons;
d. The message is in the wrong 'language';
e. The medium(s) used.

It is the regulation of the boundaries, the minimisation of the influencing factors and the software systems used that control the level of reporting complexity in projects.

The level of complexity increases:

a. When requests from various parties are not coordinated or standardised;
b. The frequency of reporting does not fit the systems available, or the timetable;
c. The systems used do not minimise / reduce duplication of effort;
d. The boundaries are not regulated;
e. The interconnections and their critical factors are not appreciated nor understood.

You will have to accept that there are a number of stakeholders at various levels that need to be informed about the project's progress. You as project control will need to make sure that the information is timely, accurate, concise, adapted for the right audience, free of 'filtering / noise' and serving above all the project team. I will come back to the issue of 'filtering / noise' in reporting below as well as in the next chapter. I will discuss this under 'Issues' that you need to be aware of and manage. This topic is one of the major interfaces between project control and all the other disciplines / team members.

Without your team / the project team you would not have an acceptable output. Remember the RACI, the PM is accountable for what is reported from the project and if you don't have her/him on board then you have lost the battle and the war.

So, in this chapter we will consider this and I will try to guide you through the pitfalls. I will try to cover all the angles, but above all I want to give some guidance on the subject by discussing, 'for whom', 'why', 'how', 'layering the reporting levels', 'identifying the requirements', the 'reporting cycles' and the issue of 'noise' in reporting / outputs.

1. The end user

You should be asking yourself this question for any and every project control output whatever your seniority level, junior or senior. As project control you receive, question, update and report on project information. The basic context here is that if you fail to do that then you are planning to fail.

Also, the other main point is that you <u>should</u> <u>not</u> stop telling them what you are likely to have repeated a number of times before. If you assume that they know something then it is likely that you will fail. And the problem is not only that you will not have your stakeholders / parties on board, but also that after a certain number of years in the profession you will walk out feeling disappointed. Also, and this might not be part of your job (depending on your level in the organisation), you cannot assume that your audience, including project management professionals, know and understand all project management basics such as Early – Late expenditure envelope, project S-curves, Earned Value and other standard definitions, etc. You will also come across some individuals that will distort standard definitions because, well 'it is not practical', 'that's a lot of theory and cannot be implemented', 'it is subjective', etc.

It is a minefield, but that's what makes life interesting and when you get to these very rare 'YES' moments you get a massive dose of satisfaction.

The next question then is: *What is the required approach?*

Since you, as the project controller, will know where the integrated information is located, then you need to present it as such and you will need to highlight this. Then you need to consult others. Do not assume that all PMs think the same way and therefore require the same report(s). Do not just consult your PM but speak to the other team members, usually working for different Line Managers or Companies. They have requirements and it is most likely that they need some reports that contain different information. Then move to the project stakeholders, who have a vested interest in the project and different sets of information (I am including here the colleagues in Finance). So, you have gathered all the reporting requirements and above all their purpose.

That is when you start integrating.

You will work based on the requirements, the information available and having understood which tools you will use and how you will deliver the requirements. You have to manage the interfaces and the stakeholders and you should not hesitate to indicate conflicts between requirements, project management standards and information available. This means that you need to highlight facts such as the necessity of exporting information from the standard software tools to others and manipulating the information provided. I am not saying that you will be successful with this, but as a professional you have to address the faults of this approach (more ammunition on this in the following chapters of this book).

Your supporting argument for not exporting to other software packages, which you will use, is the ability to drill down to lower levels or roll up to higher level of information and be able to substantiate the information you provide.

Getting your stakeholders on board will be critical and this includes the project team. However, you cannot rest on simple acceptances. There has to be continuous communication – verbal and written, about what has been agreed, tell them again and again and again. It will be advantageous, after you have gathered the requirements and you have a sample of the outputs agreed, to present the sample reports.

2. Rationale – The 'Why'

There is, unfortunately, a simple answer here, 'a picture paints a thousand words' and some people are more visual and others do not have the time to read overloaded pages of reports'.

Try to get approval for the outputs to be produced from your PM(s), the team and the other stakeholders. It will also be good to obtain approval from more senior levels, reminding them what has been agreed.

The other point that you have to consider will be that the standard project control documentation, for example the Project Control Plan (PCP), has to indicate the reporting output(s) together with the timing. So, as a discipline, your PCP is the standard that you have to adhere to.

3. How to bring information together

In thinking about the 'how' you need to consider bringing together everything that has been described earlier.

If you think about it, any information created in a project rolls up from the more detailed to the more summarised level. This then goes up another level and so on to reach the appropriate level, as per Figure 8.1 below.

Figure 8.1: D*iagram indicating the rolling up of project information.*

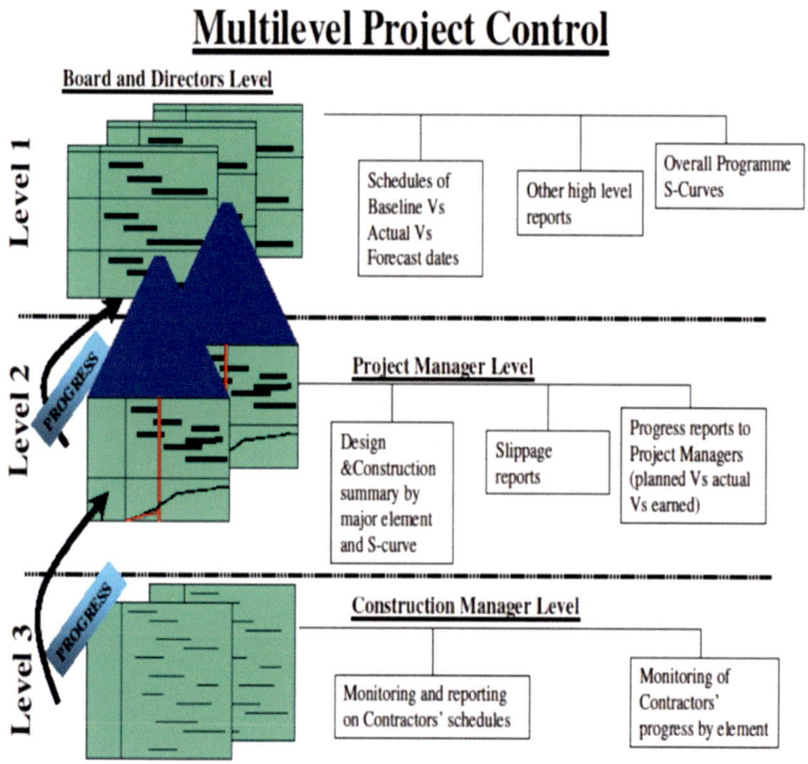

*What is the basis for rolling up information then?*

Yes, you guessed it right, the WBS and this is one of its purposes. It is about grouping and rolling up information using the project control software, which has this as a standard code.

You can imagine now what and how information on a number of projects rolls up to higher levels using the WBS.

However, I am not saying that this is the only way you can roll up project information.

You can use the CBS or the OBS, coding, etc, as long as the structures have been standardised and it is clear to your team and stakeholders that this is what you are doing.

Rolling information up through the standard project management structures and coding is very important and mitigates the risk of reporting integrity and complexity. Even if you have to export the data the standard structures will support the process and you can always get back to the source to cross check. You do however need to control the export and what happens to the data after.

Next, we will review the number of levels you can use for layering the project information and the level of detail you need to get to.

4. Layering of the output levels

Building from the project WBS upwards to programme structures you can define the levels of layers you can use for reporting / outputs. You will always have to think about your audience.

a. How many levels?

At the lowest level of the WBS – levels 4, 5 or below – you have to think of the information required by the teams that on a day by day basis work to deliver the project. This is usually expressed as task cards that the teams should be given on a weekly basis, reporting on what has been completed, what is in progress and what is coming up. In the 'Last Planner' method weekly tasks are rolled-down as well as upwards.

The levels above this are usually disciplinary ones or to the parties that contribute to the project (contractors, design consultants, etc.).

Above this we are looking at the project manager level, which obviously is focusing on the rolled up versions of the WBS but could, if and as required, be able to drill down to the lower levels.

Following on from that and using the programme of works level structure, information rolls up to the higher organisational levels.

The structure and the coding should also allow for horizontal interrogation of information for departments, such as Finance, Procurement, Logistics, etc., as well as the Contracting parties, especially in the case where delivery is done through Alliance / Partnering.

b. What level of detail?

Your level of reporting detail is dependent on the requirements, which could vary based on the resource, as in person-time information loaded on the project activities, to LPMO (Labour, Plant, Material, Other) costs, to direct or overhead costs, to project summary level, etc.

Another level of detail, usually favourable with a number of management levels is 'milestone reporting'. These are / should be embedded in the structure and could as well be coded separately.

The structures and coding should allow you to slice and dice the information according to the stakeholder's requirements. We discussed earlier how you can use the software tools available to design your project control system and therefore the reporting outputs. You should have in mind that monitoring and reporting is one of the project management success factors and therefore if the level of reported details is not adequate, project control will be considered as a failure. As an aside, this is probably why in the last 50 – 60 years project control has not been as successful as it should be.

So, structures and coding are part of the tools to use for layering project information and these allow you to roll this up to the various stakeholders. You will though identify what are the required outputs.

5. Identifying the required outputs

Identifying the required outputs is not as easy as it may first seem and we touched on this earlier. In the following pages I will take you through some of the points to consider.

a. How?

Assuming that no guidance on requirements is given and you have to identify the required reports / outputs, instead of starting with *'tell me what you need'* produce a document that identifies the standard project management outputs layered by levels of detail. For example, reports that can roll up information from activity and possibly the lowest WBS levels to project level and higher than that. Do not forget to include reports for the various disciplines that support the project. You do not want them to (a) complain and (b) go away and create their own reports (uncontrolled output).

After that you should discuss requirements with each of the interested parties. This is where you challenge, as far as possible, what you are asked to deliver against standard outputs. By that I mean if a substantial part of the requirement is covered by the standard report then you have to explain why what is available can accommodate the requested report, without additional effort.

Have samples of reports available and if necessary take your audience / stakeholders through the standard report layouts on the software screen.

Having said that, not all requirements can be accommodated through standard reports. To allow for these demands you will need to consider using, in the best case, a reporting software tool, or, in the worst case, somehow standardising the export of data. The reason I am saying this is because whatever you do, security-wise, exported data will be in the best case manipulated or in the worst case corrupted and you will not know about it. A piece of perhaps obvious advice here, when you export data always date and store the exported file, you never know when you will need to refer back to it.

When briefing on reports always emphasize the use of the standard structures and do not allow for externally used structures. If your 'customer(s)' insist, then you can always say that you will not own the output.

b. Obtaining acceptance

When you are comfortable that you have agreed the requirements then you will need to get acceptance.

Do not assume that it is there.

If not in writing, which is a bit too much in a team environment, have the acceptance given in front of others. There are various means to achieve this, but you should not ignore it.

An area that you will need to focus on, in terms of acceptance, is that you, as project control, are reporting on project life-cycle whereas finance is looking at financial year ends. The importance of this is on the values and performance you are reporting against. In particular your challenge is around reporting and obtaining acceptance on Earned Value reports.

c. Following up

Now, as things move on, so do the reporting requirements. These cannot be cast in stone never to be revisited.

The basic reporting outputs remain constant but your stakeholders' requirements could change as the project progresses. You should expect that as the project moves through the various stages, from Concept to Design, Implementation and Commissioning, reporting requirements will vary. At the implementation stage, lower level detail will be required and therefore there will be more reporting demands.

You need to be prepared for the follow up reporting frequency and you might even set up a reporting requirements review timetable. This is not only good practice but it also allows you as project control to be a step ahead and ready for the discussions.

6. Setting the Reporting cycle

In every project the teams and the organisations supporting it have a reporting cycle. This is usually a period where information needs to be rolled-up monthly and should follow the financial / corporate reporting cycles.

Therefore, the project reporting cycle needs to comply with and support the various requirements in terms of deadlines. So, as project control you will need to consider for whom is the report, when is the cut-off and how you will organise the information to be available to them.

a. For whom?

As project control you know who the external contributors to your reporting are and also who is expecting your reports. That is, who are the contractors, consultants and other parties that need to provide you the information at the right time as well as the internal parties, e.g. the PM, design, finance and others.

In the case of an Alliance / Partnering environment you will need to be aware of their corporate requirements and in particular what information they are expecting to receive back. This will have to be agreed during the stage where you are establishing the reporting requirements.

b. When?

Reporting cut-offs are very interesting because you need to establish:

i. When will the parties to the project provide you the information and what information that will be. The usual practice is to receive or generate progress to the end of the reporting period, with actuals to the middle of the period and accrued values (or Value of Work Done) to the end of the period. This is where the 'contract' terms will influence what you are capturing.

ii. When you will tell the PMs that his/her input is required by and how long they have to complete their project reviews with the teams. In the case of supporting Programmes of work, you will need to consider at what stage reporting cut-offs will accommodate their requirements for information to their level and then upwards to Director level.

iii. How you will interface with the corporate tools / departments (mainly finance) so that project reporting is understood from their side.

c. How?

Well the answer to this question is straight forward, as long as you have worked out the above.

Your reporting cut-offs should be included in the Project Control Plan (PCP) as an Appendix and this will need to be communicated to all.

As the reporting cycle is mostly formed from the higher level requirements, e.g. the Board meetings and these are issued near the end of the year you will need to align the project reporting cycle to these. Whenever you revise / update the monthly reporting cycle this will have to be included in the PCP.

7. The issue of 'noise / filters' in reporting / outputs

We come now to review the issue of 'noise / filtering' in reporting and how these lead to misunderstandings and complexity.

a. Managing project communication and information

Looking at the issue in a simple way, the two figures (Figure 8.2 and Figure 8.3) below give you an indication of the simple / theoretical reporting steps and what happens in reality.

Figure 8.2: *Diagram representing the project reporting.*

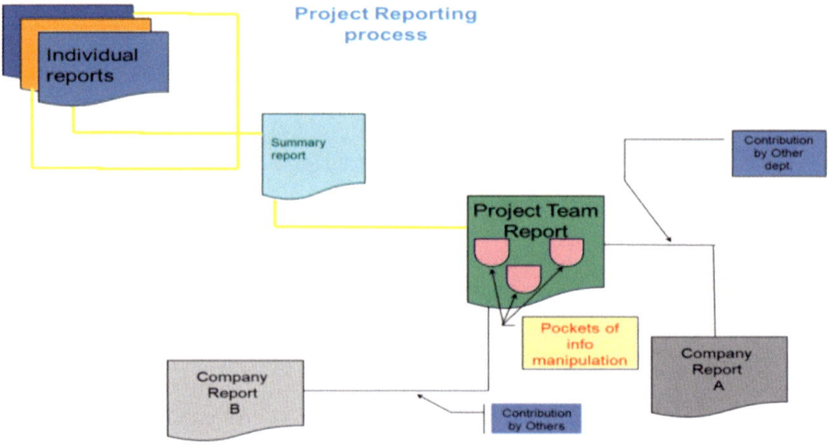

Figure 8.3: *Diagram representing the filtering of information rolling up.*

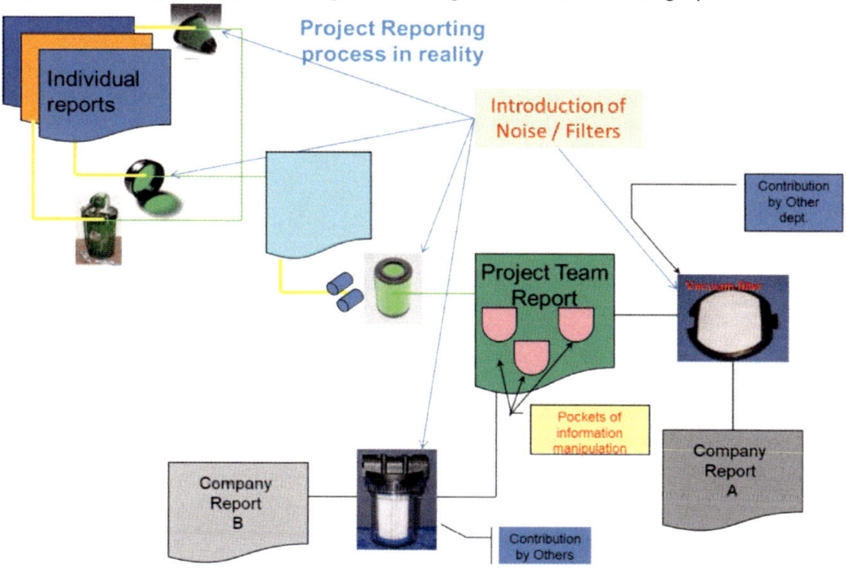


As Figure 8.3 shows, a number of 'filters / noise' levels are introduced in the process of reporting and at the various rolling up stages. In this section we will examine why this is happening, the complexity introduced and what needs to be considered.

Both BoKs (from APM and PMI) and a number of authors (Moore, 2002; Turner, 1999) highlight the importance of organisational structures and communication / information management in projects. Without a structure 'people will report progress in a way which puts them in the best light' (Turner, 1999). Reporting, as part of the data-gathering process, affects / influences the project organisational structure.

Under 'Principles of Good Project Management' Turner (1999) states that there is a requirement for clear and simple reporting structure "simple reporting at each level of the breakdown" (p. 86 – 88) and also highlights the importance of milestone reporting.

At Terminal 5 the Management Team, from the very early days, caught on to this principle and came up with the slogan *7007*, meaning 70 milestones by 2007. This gave clear indication to everyone of the importance of the 70 milestones that were critical to deliver the programme of works by 2007 (actually the expected completion of construction).

Strong messages communicated appropriately provide the necessary emphasis that needs to be conveyed internally and externally. In the case of Terminal 5, information management on the reporting side revolved around the identified milestones.

Information flow is bi-directional (Ford & Sterman, 2005) and not uni-directional (Pritchard, 2004) and (Frame, 2002) and therefore the appropriate formula to be used in projects should be ($n^2$-n), where 'n' is the number of people involved - information nodes. The importance here is that for each interconnection, for and from each person, or communication node, there are different variables that need to be considered. These variables can be seen in the following definitions, as given by Moore (2002) and Prichard (2004):

- Encoder = sender;
- Decoder = receiver;
- Filters that are used by both the Encoder & Decoder;
- The message;
- The medium (what is used for transfer the message).

Each bi-directional communication line carries and possibly amplifies at least the last three variables. Further on, Prichard (2004) classifies the number of communication problems that one could come across under three categories, a) sender or receiver, b) message and c) medium problems. Some of those problems are listed below and I will provide an indication of a possible outcome.

- Sender / receiver problems: "The message is received in a format that is not understood" (Prichard, 2004); therefore complexity and confusion arises, or "The message is received but misinterpreted" (Prichard, 2004); resulting in increased complexity;
- Message problem: "The message is in the wrong language" (Prichard, 2004), e.g. jargon used by sender; therefore again confusion and complexity arises;
- Medium problem: "The wrong medium is chosen" (Prichard, 2004), e.g. the PM sends a sensitive message about the customer relationship via email; therefore complexity arises.

Prichard (2004) points out that for communication purposes, "the information from project management software tools needs to be transferred / transferable to other tools and applications including spreadsheet", etc.. However, Jaafari and Manivong (1998) indicate this as a being problematic. The transfer of information and the required manipulation increases the workload, the number of interfaces, the likelihood of misinterpretation / misrepresentation and since there are more interconnections, we have increased complexity.

b. Complexity

Since communication / information flow interconnects the project sub-systems and the number of the parties, it should also be the carrier of complexity. It is not uncommon in projects for reports to be interpreted by decoders in a different way (Turner, 1999), or as different levels within the participating organisations.

Individuals do not receive the required information, or they use other means to produce their reports. As the reconfigured information flows in different directions and not from an appropriate source it introduces further complexity.

Complexity also arises as the flow of information between the different systems and sub-systems is not appropriately encoded, or the wrong information is captured and encoded. This is even more significant when considerable time elapses from the commencement of an activity, contract, etc. to reporting / encoding the information in an appropriate manner.

Reporting / encoding causes uncertainty and concern within the project team which results in self-organisation and attempting to take appropriate action. It is profound that as a complex system the project utilises self-organisation to overcome difficulties. Frequently project teams, in order to avoid the consequences of not reporting, are busily creating their own reporting mechanisms and monitoring tools, or else described as entropy (Moore, 2002). As other forms of reports and communications are imposed on them, which is an attempt to impose order or negentropy, it is perceived by the team as a restrictive approach often resulting in conflict.

We will examine complexity in reporting from the point of view of the encoding and decoding processes and the way information is prepared and received.

Figure 8.4 below reflects the context of reporting only at the high level. During encoding and decoding a number of factors influence the output (Prichard, 2004). Thus, reports can carry a mixture of errors and error adjustments (as shown in Figure 8.4). As the message is transferred through the interfaces it is affected by these factors and the output in many cases does not resemble the input.

In Figure 8.4, the Project Manager during the encoding process gathers information from the number of reports (rep1,2,3,4,...) that have been issued by the various parties to the project. The start of the report writing process has been preceded by the progress meeting, where issues have been raised, decisions and actions have been agreed in the minutes and all participants have left the meeting with various different messages, feelings, and thoughts on outputs.

The encoder 'filters' information / messages based on verbal information received or outputs in different formats and medium. If the approach has not been defined from the beginning then as the output rolls up decisions taken are not only affected by behaviours that prevailed during the meeting but also using 'filtered' outputs.

Complexity arises because the 'interactions of the structures that are important' (Lucas, 2000a) are affected by a number of factors and encoded errors. It should be noted that even though information is carried mainly through reports, further interactions are occurring from the day-to-day operations when direct feedback is given to the parties of the project. These are highly likely to become filters which are used by the encoders, decoders or other recipients of the reports (stakeholders, e.g. project board members). Thus, project reports become the carriers / agents of propagating complexity. Additionally, the introduction of subtle encoding errors in reports that sometimes are used consciously to conceal issues (Turner, 1999), resulting also from the systems / tools available, could cause even more significant differences in interpretation.

The level of reporting complexity in projects is controlled by the regulation of the boundaries, the minimisation of influencing factors and the software systems used.

The diagram in Figure 8.4 presents the continuous process of Reporting. It indicates the interfaces, the processes and the outputs from the two reporting cycles, the PMs and the Decoder's. The factors and the encoded errors are indicated and a brief explanation is given on what is carried by the errors.

We can deduce that the level of complexity increases:

- when requests from various parties are not coordinated nor standardised;
- the frequency of reporting does not fit the systems available, nor the timetable;
- the systems used do not minimise or reduce duplication of effort;
- the boundaries are not regulated;
- the interconnections and their critical factors are not appreciated, or understood.

c. Conceptual solution for resolving the issue of 'filtering / noise' / complexity

If we could expand pictorially Figure 8.4 to include the whole flow and the considerable amount of information in projects it would have been the most chaotic illustration. This is also confirmed by Laufer (1996).

Considering reporting as a process (or sub-process), we can use the concept of the boundary/interface as the 'tool' that regulates the flow of inputs and outputs.

Therefore, the reporting process can be regulated in terms of internal and external imports (other reports), as well as exports (reports to the decoder). With the former we are attempting to regulate how and what information parties provide to the project. This is where the software tools and the structures that hold and integrate the information play a very important role on both sides. The latter in its majority needs to be understood by the team, especially the information required for the 'monthly board' reports and usually the software output and structures.

Figure 8.4: *The continuous process of Reporting.*

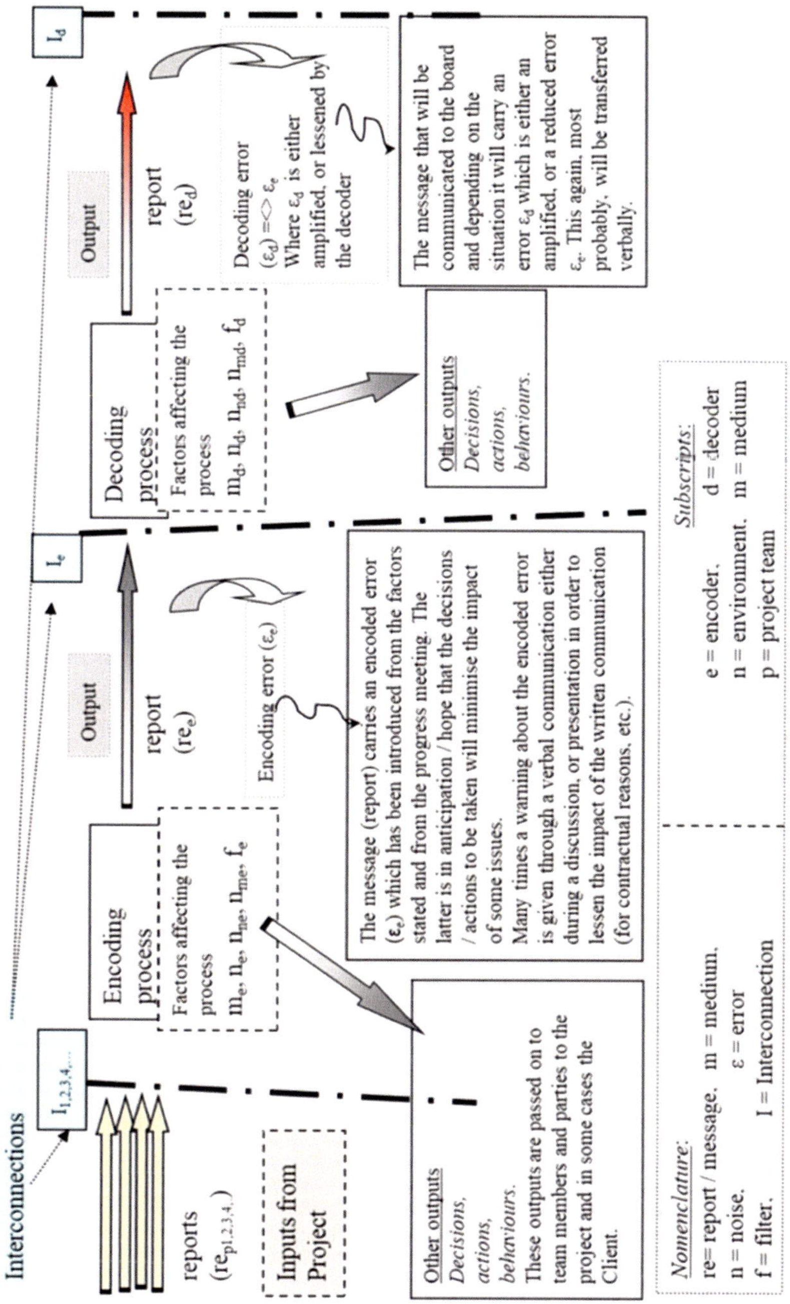

The concept of equifinality states that "there is seldom one way to achieve the required outcome". Therefore, for the reporting process the team should consider carefully how best the report will fit the project environment and how the participating individuals/parties will contribute. There must be considerable flexibility and understanding of the requirements in order to generate appropriate reports and not just conform to bureaucratic procedures.

The project team must define the communication modes through the Project Execution Plan (PEP) (Gareis & Huemann, 2000), the Project Control Plan (PCP) (Antoniadis, 2004) and the Communication Plan. The PEP must define the basic reports and the PCP must describe the processes and the details of how information is structured, integrated and rolled-up; what and how and when it is recorded, monitored and reported. The Communication Plan should be an integral part of the PCP in order to impress upon the team and the interested parties the importance of the integration of the information, the structure as well as the commonality of the tools to be used. The purpose of the two documents is to coordinate the structuring of information, minimise the transmission of inconsistent information and conflicting data formats, diminish duplication of effort and reduce the effect of the factors influencing the reporting process.

Integration of information between the various parties to the project is achieved by encoding the reporting structures which, above all must include the basic Project Management structures, e.g. WBS, OBS and CBS. All the structures should be listed in the PCP.

As projects come under time pressure, concurrency increases, thus we have an increase in complexity. The Project Manager, apart from a coordinating role, has to take a more active role in terms of managing the boundaries and ensuring consistency and frequency of good communication. Therefore, reporting has to vary according to the prevailing conditions. It does not have to be a 'once a month' activity. The number of software tools used to gather information and produce report will have to be reduced; this will reduce noise and complexity.

The PM has to define not just what, as per BoK's, but also how progress information is to be communicated and to what level. S/he should identity what could be contributing to complexity and therefore seek to minimise the propagation of error. They will have to understand filtering and remove any incorrect / erroneous data which will act as incumbent pathogens. The use of project structures and the appropriate tools will prevent the generation of bespoke information and thus minimise complexity.

The decoder's requirements, in terms of reporting, should be explored by carrying out stakeholder analysis, and using power vs level of interest and power vs level of influence matrices.

Above all, it should be accepted that reporting is not just a contractual obligation but that it is also about trust, truth, patience, understanding and not over-reacting.

d. The challenge for the future

With all the points reviewed and discussed above we have considered issues around reporting, complexity, and projects.

In this section we will seek to challenge the established thinking about reporting.

We will look at the alternative thinking and pose some questions for future consideration.

We have a powerful deduction from Niels Bohr's findings about the observer and the observed and how these cannot be independent, or as Bertelsen (2004) postulates, "they interact, and the observed system can thus not be considered as ordered in its own right. The act of observing it disturbs the order the system might have possessed before the observation". If we apply this deduction to the project environment and consider reporting as an observation upon the system / project, we could raise the following question.

Is it possible that, by requesting reports, the observer – the Project Manager, the client, or anyone that requests reports – interferes with the observed and therefore, as Bohr described, 'the observer disrupts the order of the system'?

Taking this hypothesis further, and as a consequence of being observed, the system / project responds with an increase in entropy (Moore, 2002) over reacting and reporting full activity. The observed system takes the option of 'not guilty' in the prisoners' dilemma (Axlerod, 1984) because the project / team knows very well that the bearer of bad news always suffers the consequences. To interpret this in Bohr's words 'the act of observing it disturbs the order the system might have possessed'. Thus, the action of reporting increases complexity not only in the processes but also in the behaviour of the observed. In fact the company Martin-Lockheed is reviewing the way reporting is carried out in order to minimise the effect on the observed.

Based on the above then we could ask the following questions.

Should we to look to a project world where there will be no need for reporting?

Could we also take this even further and imagine we let the people / teams do the work without any observations? Are we, the clients, the PMs, the teams, ready to accept this?

These are some of the probing questions that could be asked and researched upon. As Arthur C Clark said: "*the only way to find the limits of the possible is to go way beyond into the impossible*".

So, in summary, this is what can be done to manage the issues from the outputs / reporting:

a. Define properly the standard project management Structures;
b. Identify the boundaries;
c. Manage the boundaries using the tools and ensure consistency and frequency of good communication;
d. Describe what is required - use the tools PEP and PCP;
e. When building up the PCP, allow for flexibility, do not assume that every project is the same;
f. Reduce the number of different tools (avoid spreadsheets);
g. Reduce duplication of effort in reporting;
h. Adjust the frequency of reporting to the requirements;
i. Identify the stakeholder requirements and provide training on reading the report(s);
j. Finally, reporting is not just a contractual obligation. It is about the 5 Ts: 'Trust, Truth, Tolerance, Thoughtfulness and not going over the Top'.

In the next chapter we will revisit the complexity in reporting with some more in-depth discussion.

# 9. PROJECT CONTROL AND ISSUES TO BE CONSIDERED

In this chapter we will look at the issues to be considered when performing the processes of project control.
In the earlier chapters we looked at the issues between the functions of project control. However, there are other issues to be considered and these are around interfaces with other departments, the cultural issues, those that are generated by people judging about 'best practice' and the complexity generated by the main project control output – reporting.
I cannot say that what will be included here is a definitive list of issues, however these are certainly ones that cause major problems.

1. Project control and the interfaces with 'others'

As project control, your role is to interface with the various disciplines within the project when talking about project level control, the various departments when talking about group of projects and finally higher levels, e.g. Directorates when you are working at corporate level.

If you have not done this already then it is time to look around you.

Starting from the project level, the RACI in your PCP will describe all the other disciplines that you will need to communicate with and inform. That is the easy part; the reason it is easy is because your PCP should have been agreed / accepted by the project team members after you have shared it.

The interesting part is to understand from your 'environment' who else you need to interface with.

So, let's pick up one of the departments – Finance.

You need to make sure that your finance department knows what you are doing, why you are doing it and above all when. It will not take you long to brief them, and do not assume that by talking to them once you have briefed them. As per all your communications – tell them and tell them again and again. You need to understand that your month end figures – your output – are their input. Every month they need to understand amounts accrued, to be paid, etc. They need to do their balance sheets and all other reporting that will, at the end of the day, go to the Board and the shareholders (and obviously your monthly salary!).

It is worth investing time and resource to ensure that the interface is well managed, at all times, and that there are minimal arguments and conflicts. I have worked / interfaced with one particular finance department so well that at month end we were coordinating actions to align SAP with Primavera P6. They were using P6 to report financial information, forecast, interrogate the actuals in SAP, which does not happen frequently. They supported us on the fight with accruals. They were using our project control data!

There are plenty of similar 'positive' examples of recognising and looking after interfaces with departments such as Procurement, Treasury and others.

I can hear you thinking, “can you give us a definitive list”?

Even if I could it would be wrong, because I don’t work in your company or projects and so don’t know your set up or various other intricacies. Remember, there is **no** best practice, only good practice and we will look at this further down. If I were to give you the list it would not necessarily apply to your project control environment.

However, what I can tell you is to 'look' around you and not just at the project team. Look at who is interested in the project / programme of works information and contact them, ask them what it is they want from project control, how can you help them with accurate information and when they would want this. Even if we reverse the example and say we, as usual, want information from a department, e.g. Procurement, ask them how and what is the best way for getting the required information.

Another example will be the Company Treasury department, in the case of a client organisation. As your project or programme of works has procurement requirements a lot of the suppliers will be based outside your country and therefore they will have to be paid in foreign currency. If the Treasury department knows when payments have to be made they can hedge against currency fluctuations and save your project and the company excessive risk exposure. Therefore, if you as project control provide them with the milestone payments you will be enabling them to mitigate that risk and at the same time managing closely the interface.

The basic premise here is to understand and manage your interfaces and specifically to think of:

1. With whom and how far;
2. How to address / contract them;
3. Consider who you cannot afford to ignore.

2. The cultural issues

A number of cultural issues within an organisation, but also between organisations that come together, as in the case of partnering, will need to be addressed. Each organisation has a 'personality' / an image and a number of these are described very well by Morgan (1997).

As project control and for whichever type of organisation you are working for or with, you should be armed and able to understand the 'image' and the drivers and then adjust your behaviour accordingly.

You will have to accept that cultural interfaces and consequently behaviours of your 'customers' vary and therefore you need to be able to adjust. Let's be clear here that by adjusting I **do not** necessarily mean compromise. For example, compromising on working on a WBS is not an option. In one case it took me two hours to convince a Project Director, of an over £100M project, that we had to have a WBS. An example of where you can *'compromise'* is, depending on the requirements and as long as output is not meaningless, the level of detail in the logically linked and cost loaded programme. In this case you might give in as long as they understand the limitations on the outputs.

You will be surprised how much influence culture and behaviours exert on project control. One of the reasons that Earned Value Analysis (EVA) has had implementation issues is exactly that.

Here are 10 excuses, as presented at an earlier APM (UK Association for Project Management) event, used by practitioners for not implementing Earned Value Management and from these you can establish for yourselves how culture and behaviour have affected its implementation:

- Earned value is not suitable for my type of project / industry;
- I can't get hold of accurate resource information;
- Earned value analysis is too much work. Especially at the outset;
- Earned value analysis is a substitute for good/ experienced/project managers;
- I know where my project is, I don't need earned value;
- I have sub contracted the work…so it's not my problem;
- Earned value means open book reporting;
- Earned value analysis ignores critical path;
- Earned value analysis is a bureaucratic process. It costs too much. I can't afford it;
- Projects still fail with earned value analysis.

These together with other 'soft' issues of interactions and the type and style of the approach to be taken will have to be considered. You need to consider where and when the 'soft' / people approach stops and the more robust / 'control' approach begins. For example, at the early stages of setting up the department, or initial stages of the project your soft skills need to be deployed to win as many people as possible in the departments you are interacting / interfacing with.

There is a lot that can be achieved this way, rather than using the robust / 'control' approach. However, and after a period of time, you will be the judge of whether this is the most appropriate style. In some occasions you will have to implement the more robust approach. To give you an example, after five years of a soft implementation of the project control processes I had to revert to the more robust 'control' approach and of course some people did not like this. During these five years the department put a lot of effort and took a lot of the blame for things that had not been right. So, you will need to decide when to turn things round and of course you will have to let the team know that things are going to change for those that do not comply.

There is also the interesting cultural interface between those leading the project control and those leading other departments. The usual issue is around communication because you must not forget that project control processes affect all in project delivery and these need to be communicated widely. People will likely complain that you send too many emails, notifications, flyers, that your Project Control Plan (PCP) does not cover all project control processes (in their view) that there is too much detail or there isn't enough detail. In some cases, your 'soft' people approach will clash with that of control from others and the opposite when you start raising the issues of control. There is a fine balance between the control and the soft approach and you need to identify when to change 'mode'.

Overall in my experience, the control approach very rarely brings the required results and, in most cases, ends up in conflict between the various disciplines.

We will see below how and what steps you can take to support the soft approach to project control.

How to address the above?

Addressing the above challenge on your own is not easy (depending on the size of the teams you are supporting). Therefore, you will need some allies and above all you will need to have Senior Management briefed on your plan of action.

An approach could be to create project control champions. You will need to identify and win over several individuals who will support you and the project control practices / initiatives. To use a 'translated' complexity characteristic, the champions are / will be acting as 'autonomous' agents or 'attractors' (representing the concept of project control) that will be operating as bifurcations which manage the effects of the other complexity characteristics like 'instability'. By observing behaviours, events and other signals throughout the organisation(s) they intervene and resolve the issue(s) arising.

Having said that, you will need to ensure that messages you communicate do not end up like 'Chinese' whispers. It is worth checking and making sure that the correct message has been cascaded.

Another approach would be increasing frequency and clarity of communication including the medium you will be using. Do not assume that people read their emails. Where you can, use your champions to cascade the message and always include examples. A picture does paints a thousand words, however, sometimes project control communication needs more detail. You will need to canvas your audience and send the corresponding message.

Implementation of project control is the same with implementation of change. Winning hearts and minds is part of a proper implementation of project control. You have to think that, as happens with all projects, people will move on, so even if they take with them 30% of what proper project control is then you have contributed to the overall purpose.

3. Is there a best practice project control?

Often, we hear that we have to follow 'best practice'. The issue of best practice on all the disciplines is one of the main arguments between theories. From experience we all know that there is not one best practice approach. There might be common processes and outputs that are to be implemented, but we cannot say that these apply to all the environments and all the types of project control departments.

Morgan (1997:44) in his book 'Images of Organisation' indicates why adapting the organisation to the environment – contingency theory – is the approach to be followed and he notes:

- Organisations are open systems that need careful management to satisfy and balance internal needs and to adapt to environmental circumstances;
- There is no one best way of organising. The appropriate form depends on the kind of task or environment one is dealing with;
- Management must be concerned, above all else, with achieving alignments and good fits;
- Different types or species of organizations are needed in different types of environments.

All the above apply to project control as a practice.

What the above is saying is: understand that project control needs to be considered as a 'live organism' rather than a mechanistic one. It is people that make the project control work and deliver the project, not blind compliance to the 'control' approach.

Work with and understand the 'situations' that prevail, allow for flexibility in the processes, ensure people understand, use the outputs don't let them 'use you'. Make the higher hierarchical levels understand this and make 'emergence' or simply '2 + 2 = 5' work.

After 50 – 60 years of project management theory we need to make this change of paradigm. The blinded control approach has not worked and that is why we are failing the projects and the project teams.

The best way to set up / organise the project control department depends on the environment to which the department must relate and there is definitely no standard best practice. There is only 'good' practice. This should be based on the project management theory and the basic standards that have been developed in the last 50 – 60 years, which need to be complied with and fit in the project / company environment. Your role as project control is to have a deep understanding of how these can be adapted and adopted by the project or programme of works. Just a warning here, the usual process that gets distorted is that of Earned Value and you will need to consider where to 'draw a line' and how to distinguish between the information you no longer own. It is a process that easily falls from 'good' practice to 'awful' implementation. Another process that causes issues in terms of 'good' practice is that of reporting and we are going to look into this in more detail below.

## 4. The Issues with Reporting and Complexity

A major issue to be considered is the effect of the complexity of interfaces and how this affects your output(s), amongst other things.

Project reporting and communication, at least theoretically, involves extensive investment in time from the PM and the team. For example, Prichard (2004), argues that the Project Management Institute (PMI) certification exam suggests that 90% of the PM's time is invested in communication. Such an extensive time devoted to communication highlights the need for, and importance of clear messaging and reporting during the life of the project.

Reiss (1995), Lock and Harrison (2004), Pritchard (2004), Turner (1999), collectively present the required aspects, the details for information, the manner and the content, as well as the required ability to design the project documentation and the process. They argue that the report should be made against the plan – (a common plan), have defined criteria, be issued at set intervals, discussed at formal meetings and that tools should be simple and friendly.

Pritchard (2004) indicates that reports give insight on the current standing of the project, as well as the effort taken to get it to its current stage. The literature on reporting identifies issues involving:

a) *Organisation and/or project structure and roles;*
b) *Corporate requirements*;
c) *Contract law and procedural demands*;
d) *People issues*;
e) *Software and technology intricacies*;
f) *Information / communication inadequacies.*

Each of these issues is briefly reviewed in some more detail in the subsequent sections.

a) *Structure and Roles*

The APM BoK (2000) in sections 36, 66, and 67 provides brief guidance on information management, the organisational structure, the roles and communication respectively. Whilst this guidance is at high level and easily accessible there is evidence that in practice this is not followed. Moore (2002) and Pritchard (2004) indicate that reporting, as part of the data-gathering process, affects and influences the project's organisational structure. Likewise, without structured data people will report progress in a way that puts them in the best light (Turner 1999).

b) *Corporate requirements*

Regulatory activities (Miller and Rise 1970) are concerned with information (Moore 2002). These activities in any project will have to be defined and supported by the appropriate organisational and data structure and the project information system. Also, the corporate approach to project reporting Vs financial reporting and the relevant interfaces affect project reporting. Simple establishment of protocols (Pritchard 2004) and reporting at each level of the structure (Turner 1999) should integrate information, however in many cases these serve only a bureaucratic process and nothing further. Financial 'engineering' takes over at the interfaces to replace information that comes from the project reports. This is noticed by the teams who then filter further the information from projects.

*c) Contract and Procedures*

Contractual obligations and strict procedures also make reporting seen by practitioners more like a bureaucratic process rather than a management tool to support the decision-making process. Institutions, and in some cases government departments, issue instructions and apply pressure to companies and personnel to carry out processes that on many occasions do not consider prevailing conditions, thus creating confusion and introducing / increasing conflict. The flexibility required to define the project specific documentation and the reporting process has been removed from the project management process.

d) *People issues*

The restrictions and obligations imposed raise a number of people issues, given that reporting is part of the regulatory activities (Miller and Rice 1970). This is also accentuated by the fact that management activities are frequently perceived as a non-value generating provider of services (Singo 1988). Impartiality, peer as well as management pressure and acceptance of tools (Jaafari and Manivong 1998 and Stephenson and Scrimshaw 2003) contribute to the list of problems. Turner 2005:14 states "a manager in IBM told me that junior people may avoid making honest reports in project reviews for fear of upsetting middle managers". Management and briefing of stakeholders should also be considered carefully.

*e) Software and Technology*

The use of a number of different software packages and the different approach in providing support to Project Management functions causes problems (Jaafari and Manivong 1998). This is neither effective nor efficient for the following reasons (some of them are mentioned by Jaafari and Manivong 1998 and Turner 1999):

i. Duplication of effort by a number of parties;
ii. Multiple data entries, each using their own data modelling and structure;
iii. Difficulties in coordination of information across the software packages;
iv. Lack of integrated (compounded effect) analysis;
v. Increase of complexity due to the increased number of interconnections;
vi. Reduced accuracy of transfer of realistic and/or actual information;
vii. Large number of iterations of encoding and decoding information;
viii. Manipulation of information to suit the requirements.

On many occasions it is the peoples' behaviour and acceptance of the tools and systems used that is of greater importance. Jaafari and Manivong (1998) indicate that there is a great deal of ground to be covered in terms of starting to minimise complexity with the application of technology. In addition, there are conflicting views in the current literature regarding transferability of data between software tools, with Pritchard (2004) saying it is recommended whereas Jaafari and Manivong (1998) suggesting that it should be avoided as this can increase complexity.

*f) Information / Communication*

Information, communication and the importance of interconnections in projects is also highlighted in both PMI BoK (2000) Chapter 10 and APM BoK Topic 70. Pritchard 2004. Gareis and Huemann (2000) also mention the development of a protocol of communication and other documentation that will create the links, the structure and describe the flow of information.

This flow is bi-directional (Ford and Sterman 2005) and therefore is governed by the formula ($n^2$-n), where 'n' is the number of information nodes - senders, receivers. In this bi-directional flow, Pritchard (2004) identified the following factors; filters, which are used by the encoders and decoders, the message, which is formulated by the encoders and deciphered by the decoders and the medium, usually the software or the tools used, which carries the message. Therefore, each bi-directional communication line carries and possibly amplifies or even transforms these three variables. Further on, Prichard (2004) classifies the number of communication problems that one could come across under three categories, a) encoder or decoder, b) message and c) medium problems. Some of those problems are listed below and I will provide an indication of a possible outcome.

Sender / receiver problems: "The message is received in a format that is not understood"; this results in complexity and confusion or "The message is received but misinterpreted"; resulting in increased complexity.

Message problem: "The message is in the wrong language", e.g. jargon used by sender; therefore again confusion and complexity arises.

Considering the bidirectional flow of information, the factors that influence reporting and the interdependency of the information nodes, one can immediately understand how easily complexity can arise (Williams 1999). Therefore reporting becomes an even more difficult process. Increased concurrency also increases complexity.

1. Assertions

The following examples will expand on some of the above-mentioned issues and complexities:

a) Software and Technology

Jaafari and Manivong (1998) explore a number of possibilities for creating a structure for communication, after investigating a number of options, from systems evolving through linking to intelligent agents. They establish a specific product and/or configuration model of a project as a basis for the system design, where "information sets will determine the physical nature of the end facility". A similar approach was followed with some considerable success on the construction of a major facility in UK (Antoniadis 2004). However, as the project progressed and due to other company protocols, some problems arose when attempting to monitor contingencies.

Also, it is obvious that the clarity of the message is diluted when this is not conveyed by the tool that produced it. This was identified and the practice of transferring information to and the use of spreadsheets for reporting was challenged by the Commercial Director of a major UK Organisation. On another instance, at a major programme of works, when data were transferred from the scheduling tool to a spreadsheet, complexity and confusion was generated between all the parties.

b) Structures and roles

On another major programme of works in UK, contractual requirements imposed a number of reports by 'setting' a reporting schedule. Lack of clarity, structure and roles as well as the generalisation and the attachment to contract documentation did not support the whole purpose of reporting. Within a short period following these requirements, consultants were asked to look into problems with reporting and establish proper processes and reports.

c) Corporate requirements (Project reporting Vs Other reporting)

Misrepresentation of project information is one of the major reasons for confusion and complexity, in the case of government interference / 'strategic misrepresentation' (Flyvberg 2006). The most renowned example has been that of reporting during hurricane Catrina in the United States.

At company level there is also continuous pressure on projects, at the end of the reporting year, to manipulate the information to suit the requirements – internal or external.

2. Complexity

Project management exists in a complex environment (Thompson 1967, Bertelsen 2004) therefore reporting, as a subsystem, will also exist in this environment and will be affected by the characteristics of complexity. The application of complexity theory is to enable the systematic review of the interconnections.

Lucas (2000b) states that complexity is the study of the interconnections of systems and change and that "it is the interactions of structures that are important not which objects are interacting". Reporting acts as such a means of interconnecting systems and subsystems within PM. This is achieved through the known forms of communication and can be responsible for carrying forward, concealing and in some cases increasing complexity. If we analyse and understand these characteristics we should be able to respond with the necessary actions, improve the decision making process, the output and above all improve confidence in reporting.

From the extensive work that Lucas (2000a) has carried out I list below the 16 distinct characteristics of complexity that he has identified with a brief description. I grouped these characteristics (see column 1, Table 9.1) in order to classify each characteristic for further analysis.

Table 9.1: *Complexity Characteristics based on Lucas (2000a)*

| Group | Characteristics | Brief Description |
|---|---|---|
| Conditional | Autonomous agents | Complex systems are generally composed of independent or autonomous agents (not the identical parts often assumed in science). All of these agents are regarded as equally valuable in the operation of the system |
| | Instability | Over the long term stepped evolution or catastrophes will exist (similar to punctuated equilibria). Sudden swaps between attractors become possible as the system parameters approach the boundaries of the attractors |
| | Non-equilibrium | Energy flows will drive the system away from an equilibrium position and establish semi-stable modes as dynamic attractors |
| | Non-linear | Complex system outputs are not proportional to their inputs |
| | Attractors | Self-organization relates to the presence in the system of dynamical attractors |
| Developmental | Co-evolution | The parts are regarded as evolving in conjunction with each other in order to fit into a wider system environment |
| | Self-modification | Parts can change their associations or connectivity freely - either randomly or by evolved learning procedures |
| | Self-reproduction | Usually these systems have an ability to clone identical or edited copies |
| | Downward Causation | The existence and properties of the parts themselves are affected by the emergent properties (or higher level systemic features) of the whole |
| | Mutability | Random internal changes (mutations) or innovations typically occur in these systems. |
| | Non-uniform | Each part evolves separately, giving a diversity in rule or task space |
| | Emergence | The properties of the overall system will be expected to contain functions that do not exist at part level |
| | Phase changes | Feedback processes lead to phase changes, sudden jumps in system properties |
| Behavioural | Unpredictability | In such interacting systems a chaotic sensitivity to initial conditions can occur |
| | Non-standard | ... initially homogenous systems will develop self-organizing structures dynamically |
| | Undefined values | The meaning of the system's interface with the environment is not initially specified and this must evolve |

Table 9.2 below represents selected complexity characteristics and a simple translation / description of how these can be applied to the project management environment.

Table 9.2: *Cross reference of complexity characteristics in project management*

| Characteristic | Cross referencing to project management |
|---|---|
| Autonomous agents | Representing self-organisation of individual agents, the various parties to a project, within the system, who obey different rules, or local instructions, instead of following central laws. |
| Non-standard | Demonstrating the dynamic development of self-organisation within projects. |
| Downward causation | The property that describes how the parts are constrained from the higher level features and which become boundary conditions '*on the freedom of the parts*'. Therefore, in terms of the reporting it could be easily described as the setting of the WBS, and the other project structures. |
| Self-reproduction | Systems replicate and thus copy 'errors', from one project to the next. These errors allow new system structures to evolve as well as adaptation to new boundary conditions. Therefore, there is a requirement for flexibility. |
| Undefined values | Systems that interface with the environment must evolve and thus demand the early establishment of dynamic communications. |
| Non-equilibrium, non-uniform | The later signifies the fact that complex systems, projects, evolve separately and the former the fact that influences within the system will drive it away from the equilibrium and establish dynamic attractors. |
| Self-modification | The parts, teams and members, have the flexibility to change their associations freely. |

3. Analysis

From the brief description of the complexity characteristics in Table 9.1 and the points raised under current reporting practices, a review of the issues raised was carried out. On the table below (Table 9.3) an indication is given about which characteristics affect which particular reporting issue. Table 9.3 also indicates an initial but non-exhaustive cross-linking.

Table 9.3: *Cross reference of reporting issues and complexity characteristics*

| Reporting process issue | Complexity characteristic |
|---|---|
| Organisational Structure | Affected by Self-modification, autonomous agents and instability |
| Roles within the organisations and the project teams | Affected by Mutability and instability |
| Structure of data | Affected by Downward causation and unpredictability |
| Corporate requirements | Affected by Phase changes |
| Procedural and contract law requirements | Affected by Unpredictability as the sensitivity of the initial conditions (Law, Institutions) affects the project set up |
| People issues | Affected by Co-evolution, unpredictability and non-standard |
| Software and technology | Affected by Non-linear, attractors and self-reproduction |
| Information and Communication | Affected by Attractors, phase changes and instability |
| Project characteristics e.g. monetary size and frequency of reporting | Affected by unpredictability, non-standard, undefined values, non-equilibrium.), co-evolution, non-linear (and in the case of weekly reporting it acts in a reverse way. Instead off increasing value it reduces it |

Having identified the particular complexity characteristics we will proceed to propose appropriate actions in order to improve the reporting subsystem.

## 4. Resolving Reporting Complexity

From the review and analysis, it can be seen that reporting as a process can be analysed and reviewed in the context of complexity. The aim here is to demonstrate how, by using current knowledge and understanding, we can overcome complexity that is raised due to reporting.

Communication and information flow, which generates reports, is not only coming into and flowing out of the process, but it is also created within and by the process and is affected by complexity. Figure 9.1 below gives a diagrammatic representation.

Figure 9.1: *Diagrammatic representation of the effects on reporting process.*

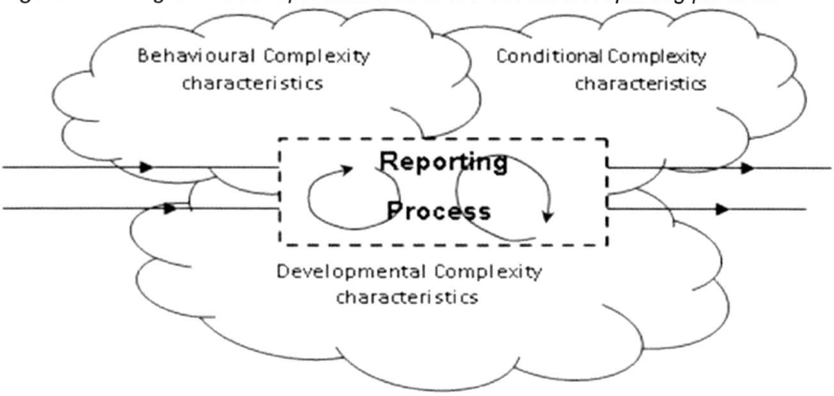


When structuring the project organisation, consideration should be given to the individual parties that contribute to the project and the agents that equally contribute to the execution of the project. Dictating and being restrictive with the output might not be appropriate because it does not permit self-modification and it will induce instability. Instead, the level of required information can be standardised by introducing the established project management structures down to a certain level.

By introducing downward causation, to a certain level, the undefined values (interfaces) are established from a very early stage and the autonomous agents are able to self-modify their corporate reports to the project requirements.

By allowing for flexibility at the lower level, the various parties, in cooperation with the project team, will be able to mutate and co-evolve, creating a linked system of information at the lowest level, where measurable information exists. From the project management side, the structured information will allow for drilling-down, or rolling-up to various levels, minimising the occurrence of filtering. It will also increase the confidence of the various stakeholders. The stakeholders will have to be introduced to and trained on the reporting structure.

The project team will also have to consider the level to which information is addressed, its density and/or level of complication. This will adjust the levels of non-equilibrium, which if not monitored tend to drive the system away from the equilibrium position. It will also adjust the level of acceptance from the parties outside the project, both client and individual organisations.

However, these energy flows vary during the life of the project, with different parties starting at different stages and changes happening to the recipients – decoders.

Therefore, the team will have to introduce dynamic attractors, experienced project support personnel, that will monitor continuously the system, make the necessary adjustments and control instability. As these attractors sense chaotic actions in their boundaries they intervene to advise and introduce stability within their own sphere.

As the project environment becomes more complex and demanding the importance of the project information system being flexible, instantaneous, comprehensive and intelligent is paramount (Jaafari and Manivong 1998). The attractors could enable such activity. The creation of a flexible and structured information system will accommodate these demands but will also remove the threat of duplication of effort, the number of different filters and levels of noise introduced from the different agents and the use of different software tools. The availability of web-based collaborative tools enables all mentioned here as well as providing the non-linearity which should exist in any reporting system.

The project team will need to prepare and discuss a Stakeholder Analysis matrix, using Power Vs Level of Interest and Power Vs Level of Influence matrices. This will give a clear indication of interconnections, boundaries and the stakeholder's required level of information. The continuous review of the stakeholders' requirements, including changes of the major players, will allow for reduction of instability due to changes in the parameters. The analysis also should include requirements dependant on the project characteristics. For example, if the project is small, e.g. 13 weeks duration and below £5M, the agents will need to define the frequency of reporting.

With regard to frequency and content, the team should examine and critically discuss with the decoder the deliverable (report) in terms of value adding, content or 'how new are the news', and level of detail. Single page reporting - dashboard approach, utilisation and necessity versus contract requirements should be considered.

All the definitions and descriptions of structures, coding, frequency and media to be used should be described in the standard project management documentation. In particular, the communication plan should be part of the Project Control Plan (Antoniadis 2004), which describes all the necessary processes as well as the interfaces within the project environment.

On the behavioural side, the agents react not only to the content but also to the way information is presented, the 'texture' of the message as well as the (management) level that has been included in the output, resulting in the occurrence of instability and conflict. Encoders will have to consider that behaviours change when things are put on paper. Project team members consider that anything on paper is a contractual action and relationships change, unpredictability and filtering increase, complexity propagates and the level of performance decreases.

Duplication of effort between autonomous agents must also be reduced. There is no worse de-motivating factor than knowing that someone else is producing the same report, that the report will not be even read, or that some form of manipulation of information will occur. In this case, self-reproduction is the biggest threat to the project.

It is also unreasonable to have personnel employed solely for chasing progress information. The message is diluted as the encoder is not the real owner / the source of the information. This type of differentiation does not contribute to the improvement of the system, or the reduction of complexity.

In a similar manner, interventions by more senior personnel should be minimised. Above all it should be accepted that reporting is not just a contractual obligation but that it is also about trust, truth, patience and understanding, and not over-reacting.

There is a requirement for flexibility in reporting, by fixing the outputs we are restricting evolution of the subsystem as the project progresses. It is the regulation of the boundaries, the minimisation of influencing factors and the software systems used that control the level of reporting complexity in projects.

We can thus deduce that the level of complexity increases when:

a) Requests from various parties are not coordinated nor standardised;
b) The frequency of reporting does not fit the systems available, nor the timetable;
c) The systems used do not minimise / reduce duplication of effort;
d) The boundaries are not regulated;
e) The interconnections and their critical factors are not appreciated, nor understood;
f) The communication channels are not kept to a manageable level.

Project Managers (PMs), who have to respond to the contractual demands should be able to review the process and take appropriate action(s). Also, institutions that enforce the process will have to reconsider the 'practice' of reporting and provide appropriate guidance.

Is there a ready-made solution that can be given to PMs so that they can overcome the problems and avoid complexity?

Equifinality teaches us that there is not a single solution to a problem, and if each project and project manager are different then there is not a simple solution.

Above all there should not be one standardised answer, because this will only cause more problems.

PMs using their experience, and with guidance from more senior personnel, should consider the project intricacies, the stakeholders' expectations (in terms of reporting), the level of comfort of the client and the other members of the project team (from the different parties) and then formulate a strategy. This strategy must be discussed with the client, the decoder, and the other parties that contribute to the generation of the information and the other encoders. This will reduce filtering, and therefore minimise complexity.

Project control plays an important part in all the above and needs to support the PM in managing these effects as well as making sure that the 'noise / filtering' is minimised. ERP and standardised reports contribute to that and therefore the driver from the project control side has to be standardisation, as much as possible, as well as the points identified above.

In the next chapter I will describe who is responsible for the project control support and what should be done.

# 10. PROJECT CONTROL AND THE SUPPORT IT PROVIDES

As Project Control plays a major supporting function either in projects, whatever the size, or companies, through the Programme Management Office (PMO), I will briefly describe below some of the main points as to whom and how.

I will not focus on the support it provides to / receives from the site / delivery teams as this is fairly major part of the responsibility and to cover it in enough detail would require a dedicated book.

1. Project / Project Manager

In different countries project control plays different roles. In the UK, project control supports the Project Manager (and the team) to deliver the project.

Your role as project control is to enable the PM, using information you are providing, to make informed decisions about the likely future scenarios. Working with the required tools and questioning information provided from other parties you should be able to offer the right level of support. By that I do not mean that you simply hand out reports but, using project control information offer advice and support on possible scenario planning, etc. Also, your performance reports have to be accurate and substantiate information given by others, who of course you have challenged already.

You should be raising your performance to such a level that will allow your PM and the team to consider you as the obvious cover in his/her absence and this is not an easy task. The reason for this is that it does not only require knowledge about the project but it also requires respect from the other team members. If you have achieved that then you can feel comfortable that the support you provide is at the right level.

2. Programme Management Office (PMO)

In many occasions, project control is part of the PMO and as such it needs to support the other functions.

It is an interesting interface, especially when the PMO role is as described in textbooks. That is, and excluding having the delivery function (PMs), one that sets up of programmes of work, carries out governance and chairing programme boards, monitors and conducts configuration management, etc. As project control you will need to interface on a daily basis with the other PMO functions and enable the integration of the project control information with those from each individual function.

You will need to understand their drivers and how your information is not only relevant and timely but is also formed of and dependent on the appropriate structure(s).

This is where points that we discussed earlier about the WBS, CBS etc. have to be communicated and followed well. It is not an easy task. Explaining with examples why these need to be followed should be your priority. You will need to start from a number of different directions and work your way towards confirmation of acceptance by all.

Having won their support then the benefits and the implementation is again all yours.

It is the only way you will be sure that the future will not be as painful.

Then the normal activities of project control start with extracting information on cost, time, progress, change control and in addition relevant governance activities. Regarding the latter, I mean supporting the PM in preparing and sometimes presenting to the Portfolio Board as well as other boards, e.g. Risk Management Board.

3. Other disciplines – Design, etc.

Linked to what I said a few lines above, is your support to the other disciplines in the project. They are the team members that will give, as well as receive, project control information from you.

To note, you should also be considering site teams who deliver these works, but, as mentioned above, this book will not cover site team support. This is a much larger topic which would need more time dedicated to it.

We discussed in earlier sections the information you require, so let's see what information you need to give / produce for the other departments or respective team members.

The various disciplines supporting the project will give and expect reports on costs, resources (as in person hours) and baseline versus current information for expenditure or output milestones.

In the case of your design team members, it is very likely that they could give you design outputs, e.g. number of drawings per activity, which they will be linking to their drawing / document register and with which they measure progress. The significance of this is that this is one of the few means of doing proper Earned Value Analysis on design activities. Another important reason is that with the advent of Building Information Modelling (BIM) and 3D / 4D etc., drawings have to link to the programme. And this is where the industry, at least in the UK, will have to pay some very serious attention and start using planning software, forgetting the spreadsheets that are killing all such initiatives.

For clarity, by design I mean all the relevant disciplines – Architects, Services, Geotechnical, etc.

Other disciplines that you need to have in mind are Procurement, for obvious reasons (without them you will not get much on site) and Consents, who will resolve any land or other issues.

So, you really need to consider all the project team members and see how you can support them. Don't forget that they are most probably reporting to a different line manager and they are responsible for that part of the budget within your project or programme of works.

4. Directorate / Senior Management / Project Director

This is the level that will receive your reports rolled up from your project(s) and obviously will be using your information in order to make decisions at the higher level.

You should have in mind what was discussed in the earlier section about 'reporting and complexity', as well as 'noise' introduced in reports. By that I want to make it clear that apart from the figures, mainly reported, there are also other factors that will influence your reporting upstream. You will need to be able to manage the message to be communicated.

The most important factor when supporting these levels is clarity. Using a multitude of supporting tools will cause data integrity issues and it is for you to stand your ground on promoting integrated information, from the project control side. Based on what was said earlier you should be able to support your case.

Understanding the need and the requirements is where you will need to spend your efforts, not only at the beginning but throughout the project lifecycle. Requirements change as the project progresses and Senior Management focus also changes. As project control you need to keep up and always think ahead: what will your customers need in the future and above all how will you accommodate this. You will also need to remember that it is not all about time and cost. Current software tools allow you to code information in many ways. So, you will need to ask about as well as propose reporting that will ease decision making at these levels.

5. The Company

Finally, you have to remember that project information is part of a bigger family of data that rolls up to company level.

Your information rolls up through your tools to corporate level. Therefore, understanding where it goes and for what purpose is one of your objectives. Discuss this with your colleagues in Finance, Treasury and Regulations (for those working in a client organisation), Procurement and others. They also have to make decisions where some of these are longer term with wider company consequences and, in the case of Treasury, involve risk.

They will be also interested in timing of this information. Your Procurement colleagues will need to know about your procurement milestones fairly early. They will also be informing you about the terms of payments, which you will need to include in your cost loaded programmes.

Overall, your audience is almost everyone around the project or programme of work you are supporting as project control. They will most probably know this but if they don't it is your responsibility to explain this to them. Above all, it is for you to convince of the absolute necessity of proper standard project management structures (WBS, CBS, etc.) as well as the criticality of logically linked and cost loaded programme.

I believe I have given you some ammunition as well as food for thought on Project Control and how it can support others.

# 11. CLOSING POINTS

With this book I have tried to bring some reality to the subject of project control. The problem is that in project control there are as many views as project professionals and this applies for both sides, yours and theirs, let alone the fact that different countries have different doctrines (as in the US). Just to make it clear, there are project control professionals that hold strong views and perhaps some of the points I raised will seem to them as trivial and their view will be 'just tell them to follow'. I have not tried to strike a balance, but to present the case of the important points that you need to follow and where you will need to hold your ground.

I emphasised the standard project management structures and spent some time on those because these are your foundations. You can adjust the style and type of project control according to the environment you are working in, but you cannot backtrack on the use of the structures. On this subject I have tried to give you as many examples as possible. Examples that I have implemented and that were part of real projects or programmes of work.

There was also emphasis on combining the structures and not just the theoretical RAM. Why the emphasis? Because even now, whilst writing this book, I have come across professionals that do not use the structures and just jump straight into solution mode. I have still to understand why this is happening, but that is a different subject altogether.

I then took on the challenge of what is considered as project control and the structure of the department(s).

I have tried to make very clear what I consider as project control. Perhaps some of you have different views but through my experience, in different industries and from different perspectives, e.g. client, contractor, consultant, I have presented my life long lessons learnt. The key is that project control deals with all things that have to do with time and cost in a project. If now you work for or are in the advantageous position of combining other areas or routines / processes within your project control then do take this opportunity. People will remember you more for your integrated and holistic approach rather than an isolationist one. As an individual it will be advisable, either as a new person in the profession or a seasoned one, for you to have this in your armoury.

On the structure of the department, I have given my views but here again you have to consider the environment you are operating in. Project control is always better if it is performed by one person and not split between two disciplines. Perhaps during your career you will discover why and where the benefits of combining cost and time under one person and one software tool lie.

We also looked at the other disciplines that could be under the department. More disciplines mean more issues, more boundaries and interfaces, which means more integration. You should always be looking to smooth interfaces and integration with other relevant disciplines. If you get the request to include other disciplines you should accept it and consider carefully how you will work on this.

I then presented the case of software tools to be used, why and how. I hope I have highlighted adequately the importance of a common tool to conduct project control. This is not only important for the project but also on a much wider perspective, which is that of information rolling up for project to programme to company level. After the failed early attempts to establish ERP systems in the mid to late 1990s it is about time we get these on board. To this end it helps that:

- We now have a more mature project management industry;
- We have software that does do what is required and from both sides, the ERP and PM;
- Concepts of integration have found their way into the various disciplines, to a certain extent;
- Other requirements, such as BIM, demand a structured approach, which then supports the implementation of ERP.

Generally, having lived and been supportive of these principles right from the beginning, I can say that it is paramount now to follow these. We are looking at a future where the fourth industrial revolution, that of IT/IS, is everywhere and it can help and support the integration of project control / management tools. Just as a hint on the future developments: there is so much that the project management industry will benefit from using the power Blockchains.

Lastly, we come to your two other challenges, outputs and issues.

I explained in these two areas what you have to bear in mind and how these lead to complexity. If you consider, and I am not supportive of this view, project control as part of a manufacturing/ production line then outputs are your final product. What your customers receive or use is an output from your operations. I presented a number of points and you should really have an open mind within the environment you are operating.

Together with other things that I have mentioned, outputs are the areas which will cause you issues. I have discussed how and why you need to manage these and how through the managing of the effects of complexity you will be able to manage the issues. You have to know what they are, where they come from and how to set an action plan to address them.

Something that I said at a presentation to an executive management meeting, relevant to rolling out Earned Value, but mainly applied to project control is that it requires: ***patience, persistence, perseverance and perspiration***

So, if you have decided to make project control your career, you will have to display a lot of these characteristics and the latter – '*perspiration*' will be true in both in physical and metaphysical senses.

Independent of your level of seniority, project control requires that you work with, set up and perform in a team environment. Project control is not an individualistic profession or set of processes. Without the team and above all respect from the team, you are fighting a losing battle. Leading from the front, or from behind, as in Lao Tzu's saying "*to lead people, walk behind them*", is a personal view but if your leadership style does not include setting up and working in and within a team environment then I am afraid you will need to reconsider your career options.
Another set of necessary attributes are honesty, trust and ownership.

I know this is the closing chapter of the book and perhaps these three words have not been addressed previously, but in dealing with basic project information – time and cost – people that you work and interface with have to know that you possess these attributes.

If you think about it you would also expect the same, since you base your outputs on the honesty, credibility and ownership of information given to you by others.

Having done the reading, you now have the opportunity and challenge of thinking about this. Have a go at the 'Exercises' I have provided in the next chapter. They will help you to consider various issues and encourage you to read other books to find solutions. I hope they will also help you in your everyday actions.

One final point, always own your output and accept errors that you might have made, be inquisitive and challenge.
Good luck and remember also that consulting with others helps the formulation of better solutions, but never lose track of the basics.

Well, this is it. You have reached the end of the book. I hope you have enjoyed reading it and above all that the real examples and the style helped you not only to understand but also see where the challenges are coming from and how to deal with these. Remember to:

*"See what no one else sees; See what everyone else chooses not to see out of fear conformity and laziness; See the whole world anew each day" (Patch Adams)*

# 12. EXERCISES

In the book, I gave you an overall view of Project Control, and in the various chapters I described what you need to be aware of in order to be able to work, manage and understand things about Project Control.

In this chapter, in order for you to benefit from what I have told you, I am giving you a set of exercises. You will need to think and tailor your response(s) to these exercises according to the organisation or the environment in which you are working. If you are working, or thinking about, a Client Project Control environment then your answers (not all of them) will be slightly different to those that are to be given if you are working in a Contractor environment, or a mega project. Therefore, you will need to think carefully and not assume that everything is the same.

I have given you seven areas / exercises, but this does not mean that you cannot ask yourself more, from what you have read in this book. In each exercise I have given you the objective(s) of each one so that at the end of the exercise you can reflect back and see how well you have responded. I would encourage you to 'challenge' yourself to answer other thoughts that might come to your mind as you work through these.

**The Exercises:**

Exercise 1: How do you go about setting up / structuring the department?
Describe the organisational structure, the departments / disciplines and the integration between them.
Objectives:

1. What do you need to consider in terms of the structure?
2. What are the parts / disciplines that will be included?
3. What they will have to do in order to integrate and how will you go about it?
4. What do you need to consider in terms of interfaces with others?

Exercise 2: What standards and processes need to be set up?
Describe the requirements and the content of what the Department / Individual will need to implement as part of the Project Control.
Objectives:

1. What are the standards you / Project Control will need to set up?
2. What are the processes you will need to implement / roll out?
3. What will you do to ensure appropriate implementation?

Exercise 3: What needs to be considered in terms of engagement with other Departments / Directorates / Contractors?
Describe the requirements for engagement with other Departments / Directorates / Contractors and Subcontractors.
Objectives:

1. What are the interfaces?
2. What actions will you take in terms of communication and interfaces?
3. What needs to be there to tackle the issues?

Exercise 4: What needs to be considered in terms of scaleability of project control?

Describe what you have to consider when implementing Project Control regarding the size and type of the organisation, the project and/or the programme of works.

Objectives:

1. What are the variables that you have to consider?
2. What level of organisation you will set up, in terms of Project Control?
3. What depth of processes will you implement?

Exercise 5: What are the Project Control software requirements?

Describe the software requirements and the approach to implementing an integrated Project Control approach.

Objectives:

1. What tools (software) will you use?
2. What interfaces you will have to enable within the tools and how?
3. What will the people / software interactions be and how?

Exercise 6: What do you need to consider when running the Project Control Department?

Describe the management approach to running the department.

Objectives:

1. What is the required management style?
2. What are the disciplines within the department and how will you go about integrating the input and the output?
3. What are your (as PC Manager) interfaces with other Line Managers / Senior Managers / Directors / Directorates?
    1. If you are not a PC practitioner:
       What are your interfaces with Project Control?
4. What do you have to consider in terms of escalating issues, resolving problems?

Exercise 7: What is the role of Project Control operating within a Programme Management Office?

Describe the role of the Project Control within a Programme Management Office (PMO).

Objectives:

1. What are your actions, as Project Control, within the PMO?
2. What do you do to integrate with the other departments?
3. What do you do to support the PMO Manager / Head to deliver the required actions?

# REFERENCES

Antoniadis, D N (2004) Case Study: London Heathrow Terminal 3 – Pier 6 Project Control. In Pantouvakis, J P (Ed), *3rd Scientific Conference on Project Management "Clustering in Construction Project Management"*, 24-25 Sep-04, Thessaloniki, Greece. Vol.1, 346-54.

Antoniadis, D N (2009) *Managing Complexity in Project Teams*. PhD Thesis, Loughborough University, UK

Antoniadis, D N (2013) Ch. 10 Teams and Complexity: Merging Theories towards a Finite Structure. In Banerjee S (eds) *Chaos and Complexity Theory for Management*. USA. IGI Global.

APM Knowledge: Guide: 'Introduction to Project Control' (2006)
Planning, Scheduling, Monitoring and Control (2015)
A Guide to conducting Integrated Baseline Reviews (2016)

Association for Project Management, Project Management Body of Knowledge, (2000). 4ed. Miles Dixon (Ed). London: Cambridge Publishing Management.

Association for Project Management, Project Management Body of Knowledge, (2012). 6ed. The Association for Project Management.

Atkinson, R. (1999) Project Management: Cost, Time and Quality, Two best guesses and a phenomenon, it's time to accept other success criteria. *International Journal of Project Management* Vol 17(6). pp. 337-342

Axlerod, R (1984) The Evolution of Cooperation, Princeton University Press and in Nash Equilibrium - Prisoner's dilemma, in www.wikipedia.com, accessed Nov-05

Bertelsen, S (2004) Construction management in a complexity perspective, *1st International SCRI Symposium*, 30-31 March 2005, University of Salford, UK.

Busby, J.S. and Hughes, E.J. (2004) Projects, pathogens and incubation periods. *International Journal of Project Management*, Vol. 22(5), pp. 425-434

Collins, A. and Baccarini, D. (2004) Project Success – A Survey. *Journal of Construction Research* Vol 5(2), pp. 211- 231

Das, T. K. and Teng, B.S., (2001) Trust, control and risk in strategic alliances: an integrated framework. *Organisational Studies*. Vol 22(2), 251-283

Ford, D N and Sterman, J D (2005) Overcoming the 90% syndrome: Iteration Management in Concurrent Development Projects. *Concurrent Engineering: Research and Applications*, Dec05, 177 – 86.

Flyvberg, B (2006) Don't be blinded by scope. *Project*, March 2006, **18**(8), 28-29

Gareis, R and Huemann, M (2000) Project Management Competences in the Project Oriented Organisation. In R.J. Turner and S.J. Simister (eds) *Gower Handbook of Project Management*. 3 ed. London: Gower.

Hughes, W.P. (1997) MSc in Project Management lecture notes. Reading University.

Jaafari, A and Manivong, K (1998) Towards a smart project management information system. *International Journal of Project Management* (IJPM), **16** (4) 249-65.

Leifer R and Mills P K (1996) An information processing approach for deciding upon control strategies and reducing control loss in emerging organisations. *Journal of Management* **22**(1), 113 – 137

Lock, D and Harrison, F (2004) *Advanced Project Management, A structured approach*. 4ed. London: Gower.

Lucas, C., (2000a), The Philosophy of Complexity, in www.calresco / org / lucas / philos.htm, accessed February 2005.

Lucas, C. (2000b), Setting the scene – Science, Humanity and Interaction, in www.calresco.org/setting.htm, accessed February 2005.

Miller, J H and Page S E (2007) *Complex Adaptive Systems. An introduction to Computational Models of Social Life*, New Jersey: Princeton University Press

Miller, E J and Rice, A K (1970) *Systems of Organisations: The control of task and sentient boundaries*. London: Tavistock.

Morgan, G. (1997) *Images of Organisation*. Sage. London

Moore, D (2002) *Project Management: Designing Effective Organisational Structures in Construction*. Malden MA: Blackwell science.

Ning, Y (2017) Selecting client's project control strategies in person-to-organisation transactions. *International Journal of Project Management* (IJPM), Vol 35(2) 212-220

Project Management Institution, (2000), A Guide to the Project Management Book of Knowledge, 2000 Edition, Project Management Institute, Pennsylvania, USA

Pritchard, C (2004) *The Project Management Communications Toolkit*. Artech House

Reiss G (1995) *Project Management Demystified – Today's tools and techniques*. 2ed. London: E & FN Spon

Rowley, T., Behrens, D., Krackhardt, D., (2000) Redundant governance structures: an analysis of structural and relational embeddedness in the steel and semiconductor industries. *Strategic Management Journal.* Vol 21(3), 369-386

Singo, S (1988) *Non-stock Production.* Cambridge: Productivity Press.

Stephenson, P and Scrimshaw, I C (2003) Integrated Project Management information systems in construction: a case study implementation. In Bontembi, F (Ed) *2nd International Conference on Structural and Construction Engineering (ISEC-02).* 23-26 September, Rome, Italy. Balkema Vol. 3. 2245-50.

Thompson, J.D (1967) *Organisation in Action.* New York: McGraw Hill.

Turner, R J (2005) Barriers to innovation and training. *Project Management Today.*

Turner, R J (1999) *The handbook of Project-Based Management – Improving the processes for achieving strategic objectives.* 2ed. London: McGrawHill.

Williams, T.M (1999) The need for new paradigms for complex projects. *International Journal of Project Management* (IJPM), **17** (5), 269 – 73.

# BIBLIOGRAPHY

Boyatzis, R.E. (1982) *The competent manager. A model for effective performance*. Willey. New York

Construction Industry Council (2005) Selecting the Team. Published by the Construction Industry Council. London

Cavanagh, M. (2012) *Second Order Project Management*. England. Gower

Dalcher, D. (2018) *The Evolution of Project Management Practice. From Programmes and Contracts to Benefits and Change*. London. Routledge

Davidson Frame, J. (2002) *The new Project Management. Tools for an age of rapid change, complexity and other business realities*. Second edition, 2002 Jossey-Bass

Egan, Sir J. (1998) *Rethinking Construction*, The Report of the Construction Task Force to the Deputy Prime Minister, John Prescott, on the scope for improving the quality and efficiency of UK construction, chaired by Sir John Egan (The Egan Report)

Fiedler, F.E. (1967) *A theory of leadership Effectiveness*, McGraw-Hill

Galbraith, J. (1973) *Designing Complex Organisations*. Addison Welsey. Massachusetts

Gardner, H. (2006) *Five minds for the future*. Harvard Business School Press. Boston, MA

Green, S. (2006) *The management of projects in the construction industry: context, discourse and self-identity*. In: Hodgson D, Cicmil S, editors. Making Projects Critical. Palgrave Macmillan. Hampshire

Handy, C. (1993) *Understanding Organisations*. 4th Edn. Penguin. London

Hock, D. (1999) *Birth of the Chaordic age*. Berrett-Koehler, San Francisco, CA

Hofstede, G. (1980) *Culture's Consequences*. Sage. London

Latham, M. (1994) *Constructing the Team*: Joint Review of Procurement and Contractual Arrangements in the UK Construction Industry, Department of the Environment, London: HMSO

Lawrence, P.R. and Lorch, J.W. (1967) New Management job: The Integrator. *Harvard Business Review*, Nov-Dec, pp. 142-151

Liker, J.K. (2004) *The Toyota Way. 14 Management Principles from the World's Greatest Manufacturer*. McGraw-Hill. New York

Madsen, S. (2015) *The Power of Project Leadership*. London. Kogan Page

Maylor, H. (1996) *Project Management*. Pitman. London

Mintzberg, H. (1973) *The Nature of Managerial Work*. New York. Harper and Row

Mintzberg, H. (1979) The structuring of organisations. Prentice Hall. London

Morris, P.W.G. (1972) *A study of selected building projects in the context of theories of organisation*. PhD Thesis, Department of Building, UMIST, Manchester

Morris, P.W.G. (1994) *The Management of Projects*. Thomas Telford. London

Osburn, J.D., Moran, I., Musselwhite, E. and Senger, J.H. (1990) *Self-directed work teams: The American challenge*. Homewood, I L; Business One Irwin

Partington, D. (2003) *Managing and Leading*. In People in Project Management. (Ed) Turner, J.R. Gower. Aldershot. England. pp.83-97

Reid, T. (2003) *Managing teams: the reality of life*. In Turner, R.J. (Ed), People in Project Management, Gower Publishing, Aldershot

Senge, P. (1990) *The Fifth Discipline*. The Art & Practice of The Learning Organisation. Random House, London

Senge, P., Kleiner, A., Roberts, C., Ross, R., Roth, G. and Smith, B. (1999). *The Dance of Change. The challenges of sustaining momentum in learning organisations*. Nicholas Brearly publishing; London

Slevin, D.P. and Pinto, J.K. (2004) *An overview of behavioural issues in project management*. In Morris, P W G and Pinto, J K (Eds) The Willey guide to managing projects. John Willey & Sons, Hoboken, New Jersey, p.67-85

Stacey, R.D. (2001) *Complex responsive processes in organisations. Learning and Knowledge Creation*. Routledge. London

Stacey, R.D. (2003) *Strategic Management and Organisational Dynamics: The Challenge of Complexity*. Harlow: FT/Prentice Hall

Stacey, R.D., Griffin, D. and Shaw, P. (2002) *Complexity and Management: Fad or Radical Challenge to Systems Thinking*. London. Routledge

Thompson, J.D. (1967) *Organisation in action*. McGraw Hill, New York

Thorpe, A., Edum-Fotwe, F.T. and Mead, S. (1998). *Managing construction projects within emerging information-driven business environments*. In: Fahlstedt, K. (ed), Construction and the Environment, Proceedings of the CIB World Congress, Symposium D: Managing for Sustainability- Endurance Through Change, 7-12 June 1998, Gavle, Sweden, pp.1901-1910

Walker, A. (1996) *Project Management in Construction.* 3rd Edn, Blackwell Science. Oxford

Weick, K.E. (1979) *The Social Psychology of Organizing.* Addison-Wesley. Reading

Wellins, R.S., Byham, W.C. and Wilson, J.M. (1991) *Empowered teams; Creating self-directed work groups that improve quality, productivity, and participation.* San Francisco

Williams, T.M. (2002) Modelling Complex Projects. John Willey & Sons, West Sussex

Winter, M. and Smith, C. (2006) *Rethinking Project Management*, Final Report, EPSRC Network 2004 – 2006

Yoram Wind, J. and Main, J. (1998) *Driving Change. How best companies are preparing for the 21st Century.* Kogan Page. London

# INDEX

## D



## E



## F



## G



## H

## P



## Q



## R

## S



## T

## U



## V



## W